LASER COMMUNICATION SYSTEMS

WILEY SERIES IN PURE AND APPLIED OPTICS

Advisory Editor
STANLEY S. BALLARD University of Florida

Lasers, BELA A. LENGYEL
Ultraviolet Radiation, second edition, LEWIS R. KOLLER
Introduction to Laser Physics, BELA A. LENGYEL
Laser Receivers, MONTE ROSS
The Middle Ultraviolet: its Science and Technology, A. E. S. GREEN, *Editor*
Optical Lasers in Electronics, EARL L. STEELE
Applied Optics, A Guide to Optical System Design/Volume 1, LEO LEVI
Laser Parameter Measurements Handbook, HARRY G. HEARD
Gas Lasers, ARNOLD L. BLOOM
Advanced Optical Techniques, A. C. S. VAN HEEL, *Editor*
Infrared System Engineering, R. D. HUDSON
Laser Communication Systems, WILLIAM K. PRATT
Far Infrared Spectroscopy, K. D. MOELLER and W. ROTHSCHILD (In Press)
Optical Data Processing, A. R. SHULMAN (In Press)

WILLIAM K. PRATT
University of Southern California
Los Angeles, California

LASER COMMUNICATION SYSTEMS

JOHN WILEY & SONS, INC.
NEW YORK   LONDON   SYDNEY   TORONTO

DEDICATED TO MY WIFE,
MY CHILDREN, AND MY PARENTS

ACKNOWLEDGEMENTS

I wish to acknowledge my association with Mr. K. L. Brinkman of Hughes Aircraft Company who encouraged me to work on a study of laser communications which began less than eighteen months after the development of the laser. Mr. L. S. Stokes and Mr. E. J. Vourgourakis of Hughes Aircraft Company provided valuable commentary on the subject of laser communications. Technical suggestions and editing were offered by Mr. R. J. Norton, Mr. V. L. Rideout, and Mr. G. W. Thompson of the University of Southern California. Also, I wish to express my appreciation to Professor Zohrab A. Kaprielian of the University of Southern California for his support during the completion of the book.

Chapter 13 of this book is based upon work performed by Mr. K. L. Brinkman, Mr. L. S. Stokes, and the author for Hughes Aircraft Company. The work was sponsored by the National Aeronautics and Space Administration under contracts NAS 9–879 and NAS 12–566.

PREFACE

Laser communications is a new and fast growing field which has formed from an intersection of the topics of optics and communication theory. This book is directed toward practicing physicists, electrical engineers, and their student counterparts who are interested in the field of laser communications but are perhaps acquainted with only one of the background technologies.

The text is concerned primarily with the system aspects of laser communication system design. Descriptions of system components and quantitative references to their capabilities have been kept brief intentionally since new devices and components still are being rapidly developed. Emphasis has been placed on the basic operating principles of system components with regard to the performance limitations they place on the communication system.

The first chapter presents a derivation of the basic communication range equation which links the generation of a laser carrier to the communication channel and the eventual detection of a received signal. A statistical model of laser emission and the photodetection process is developed as a basis for the subsequent analysis of communication receivers.

Chapter 2 contains a discussion of modulation and detection methods. The frequency spectrum of a modulated laser carrier is determined for various analog, pulse, and digital forms of modulation.

Chapters 3, 4, and 5 are concerned with the components of laser communication systems. The operating principles of lasers, optical modulators, photodetectors, optical antennas, and other system components are explained.

Design information is given in Chapter 6 for determining the amount of background radiation from the sun, stars, and planets which is incident upon an optical receiver. Problems associated with transmission through the earth's atmosphere are considered in Chapter 7.

Photodetection noise is analyzed in Chapter 8. In Chapter 9 the emission statistics of photodetectors are derived, and several signal detection methods are investigated.

Chapter 10 describes the operation of optical communication receivers. The noise power spectral density and signal-to-noise ratio are found at various points in the receivers. With the previous chapters as a necessary back-drop, Chapters 11 and 12 discuss the operation and performance of several important laser communication systems. Finally, Chapter 13 presents a methodological approach to the design of laser communication systems.

WILLIAM K. PRATT
Malibu, California
September 1968

CONTENTS

PHYSICAL CONSTANTS

Velocity of light: $c = 2.998 \times 10^8 \dfrac{\text{meter}}{\text{sec}}$

Planck's constant: $h = 6.624 \times 10^{-34}$ joule sec

Boltzmann's constant: $k = 1.38 \times 10^{-23} \dfrac{\text{joule}}{\text{degree Kelvin}}$

Electronic charge: $q = 1.6 \times 10^{-19}$ coulomb

$hc = 1.986 \times 10^{-23}$ joule

LASER COMMUNICATION SYSTEMS

chapter 1

INTRODUCTION

Optical communications began with the crude fire signals of early cavemen and progressed little in principle to the blinker systems of twentieth-century ships. The introduction of the laser in 1960 provided not only a more intense source of light for optical communications but also made feasible more exotic forms of modulation and detection.

Today lasers and allied communication system components have been developed to such a point that laser communication systems are not only practical for many applications but also offer significant advantages over radio frequency communication. The specific advantages are a higher potential data rate and an increased antenna gain.

Optical communication systems operate at carrier frequencies in the interval of about 10^{13} to 10^{15} Hz. In this frequency band, a modulation bandwidth of 10^{12} Hz would occupy only about 0.1% of the available spectrum. There are, of course, obvious difficulties associated with constructing optical modulators and receivers to operate at such large bandwidths.

For a given transmitting antenna size, the angular width of the transmitted beam is inversely proportional to the transmission frequency and the spatial power density at the receiver is proportional to the square of the frequency. Hence it is potentially advantageous to operate at a high carrier frequency. As a practical example, the spatial power density at the receiver is one million times larger for an optical system with a 10 *centimeter* diameter antenna operating at 10^{14} Hz than for a radio system with a 10 *meter* diameter antenna operating at 10^9 Hz. In this example the optical transmitter angular beamwidth is about 30 microradians. The quite narrow beamwidths obtainable with laser communication systems place a severe burden on the acquisition and tracking system which must accurately point the transmitted beam at the receiver.

The attributes of large information bandwidths and narrow transmitted beams of laser systems compared to radio frequency communication systems are simply a result of the high frequency of optical waves. Why not then utilize a high-intensity source such as a mercury arc lamp rather than a laser? The answer lies in two major properties of laser light: narrow spectral width and coherence.

1

Emissions from many lasers have spectral widths of 0.1 Å or less, whereas the natural line width radiation from a mercury arc lamp is much larger. Most other incoherent sources have even wider frequency spectra. The narrow line width of an optical carrier is desirable since optical filters may be inserted before a photodetector to reduce the effects of optical background radiation from the sun, stars, and other sources.

Coherence of an optical wave is the property that a wave is in phase with itself after some time interval (time coherence) and in phase at time points in space (spatial coherence). Both time coherence and spatial coherence are required of an optical wave for photomixing in optical heterodyne and homodyne receivers. The coherence property is also necessary to focus a laser beam to its diffraction limit for minimum beam divergence.

A study of laser communication systems logically begins with the establishment of physical and statistical system models. Such models form a framework for the application of analytical methods to the design of communication systems; they also reveal the physical and functional operations that must be performed within the system.

1.1 COMMUNICATION SYSTEM PHYSICAL MODEL

Figure 1-1 illustrates a physical model of a laser communication system. An information signal is coded to the modulation format (pulse-code modulation, pulse-position modulation, etc.) and applied to a modulator driver which provides the proper excitation for the modulator. The amplitude, intensity, frequency, phase, or polarization of the laser beam is modulated either by an external modulator as shown or by a modulator internal to the laser cavity. The modulated laser beam is collimated (made parallel) by the transmitter optical antenna. Signal energy gathered by the receiver optical antenna is focused onto an optical receiver. The output of the optical receiver is an electrical signal which is fed to a radio receiver for electrical signal detection operations. Frequency conversion to baseband is also performed in the radio receiver for subcarrier and heterodyne communication systems.

In a communication system the relationship between the transmitted and received signal power is described by the communication range equation. This equation provides a characterization of propagation in the communication channel, free space propagation loss, and attenuation losses in the system components.

Carrier energy losses in the modulator and transmitter optical antenna are described by the transmitter system transmissivity, τ_t, which is given by

$$\tau_t = \frac{P_A}{P_L} \tag{1-1}$$

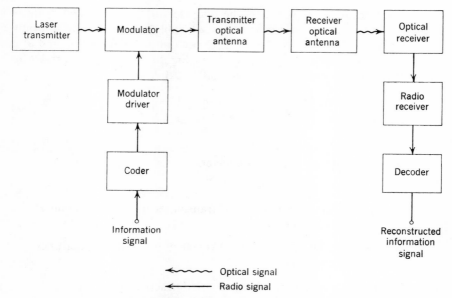

Figure 1-1 *Physical model of a laser communication system*

where P_L is the laser power and P_A is the transmitter system power. This definition of transmitter transmissivity implicitly includes any losses of beam energy due to beam spillover in the modulator or transmitter antenna.

Figure 1-2 illustrates a typical configuration of a transmitter optical antenna system which is designed to produce a collimated beam of circular cross section. Diffraction causes beam spreading in the far field inversely proportional to the aperture diameter of the transmitter optical antenna. At a

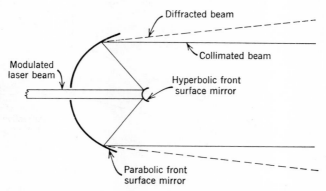

Figure 1-2 *Cross sectional view of a typical transmitter optical antenna*

distance far from the transmitter, the diameter of the collimated beam is negligible in comparison with the diffracted beam pattern. For a uniformly illuminated circular aperture the intensity per unit solid angle centered at a point, P, in the receiver plane, as shown in Figure 1-3, can be expressed in terms of the Bessel function of the first order, $J_1[\cdot]$, as

$$\mathcal{J}(P) = \left[\frac{2J_1(\pi d_T \alpha/\lambda_c)}{\pi d_T \alpha/\lambda_c}\right]^2 \mathcal{J}(0) \tag{1-2}$$

where:

d_T = transmitter aperture diameter

λ_c = transmission wavelength

α = half angle from center of transmitter aperture to point P referenced to optical axis

$\mathcal{J}(0) = \pi d_T^2 P_A/4\lambda_c^2$ = intensity at center of diffraction pattern per unit solid angle

Figure 1-4 is a plot of the circular aperture diffraction pattern [1-1].

The power incident upon the receiver plane, $P(\alpha')$, about the center of the diffraction pattern is the spatial integral of $\mathcal{J}(P)$ over a reception angle α' multiplied by the atmospheric transmissivity factor, τ_a, to account for attenuation losses in the atmosphere. Thus

$$P(\alpha') = \tau_a \mathcal{J}(0) \int_0^{2\pi} \int_0^{\alpha'} \left[\frac{2J_1(\pi d_T \alpha/\lambda_c)}{\pi d_T \alpha/\lambda_c}\right]^2 \alpha \, d\alpha \, d\psi \tag{1-3}$$

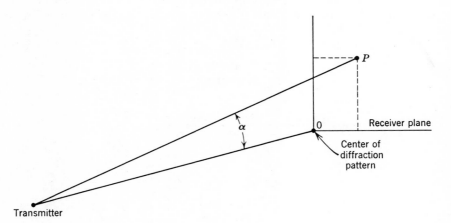

Figure 1-3   *Geometry of transmitter antenna diffraction pattern*

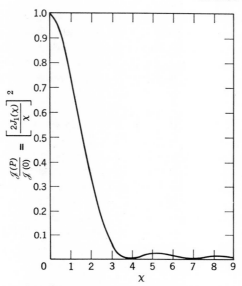

Figure 1-4   *Circular aperture diffraction pattern*

The integrals may be evaluated in terms of Bessel functions as

$$P(\alpha') = \tau_a P_A \left[1 - J_0^2\left(\frac{\pi d_T \alpha'}{\lambda_c}\right) - J_1^2\left(\frac{\pi d_T \alpha'}{\lambda_c}\right) \right] \qquad (1\text{-}4)$$

Figure 1-5, a plot of $1 - J_0^2(\chi) - J_1^2(\chi)$, represents the fraction of total transmitted energy contained in the diffraction pattern as a function of

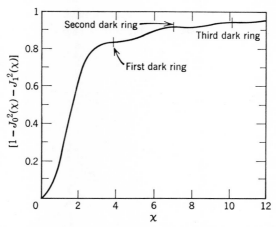

Figure 1-5   *Fraction of laser power in circle about diffraction pattern center*

distance from the pattern center [1-1]. If the receiver optical antenna of diameter d_R is located at a range R from the transmitter and centered on the optical axis, the diffraction angle is $\alpha' \approx d_R/2R$. For R large the power density at the receiver plane is nearly constant at a maximum value of $\mathscr{J}(0)$ over the receiver aperture, and the maximum value of the receiver power, $(P_R)_{max}$, is

$$(P_R)_{max} = \tau_a \mathscr{J}(0) \int_0^{2\pi} \int_0^{d_R/2R} \alpha \, d\alpha \, d\psi = \frac{\pi^2 \tau_a P_A d_T{}^2 d_R{}^2}{16R^2 \lambda_c{}^2} \qquad (1-5)$$

This is, of course, an optimistic result since pointing errors of both the transmitter and receiver antennas result in operation at some point away from the peak of the diffraction pattern. A lower bound estimate of the received power may be obtained by defining (Figure 1-6) a fictitious transmitter beamwidth as the angular diameter, $\theta_T = 2\alpha$, at the half power points of the far-field pattern. Letting $\mathscr{J}(P) = \frac{1}{2}\mathscr{J}(0)$, the argument of the Bessel function, by numerical techniques, is found to be

$$\frac{\pi d_T \theta_T}{2\lambda_c} = 1.62 \qquad (1-6)$$

and the transmitter beamwidth full angle is

$$\theta_T = 1.03 \frac{\lambda_c}{d_T} \approx \frac{\lambda_c}{d_T} \qquad (1-7)$$

Note that the transmitter beamwidth is much smaller than the angular distance between the first zeros of the diffraction pattern which is $2.44\lambda_c/d_T$.† This definition of transmitter beamwidth is reasonable only if the probability that the pointing error angle, ϵ, is greater than $\theta_T/2$ is negligible. If the probability distribution of the pointing error angle is Gaussian with variance $\sigma_\epsilon{}^2$, then setting $6\sigma_\epsilon \leq \theta_T$ gives a probability of less than 0.01 that $\epsilon > \theta_T/2$.

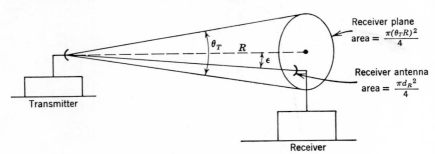

Figure 1-6   *Transmission-reception geometry*

† This is the definition of the Airy disk.

If the receiver optical antenna is illuminated by the diffraction pattern at the edge of the transmitter beamwidth where the intensity is one-half its value at the pattern center, the received power is approximately

$$(P_R)_{\min} = \tfrac{1}{2}(P_R)_{\max} = \frac{\pi^2 \tau_a P_A d_T^2 d_R^2}{32 R^2 \lambda_c^2} \tag{1-8}$$

In practice the receiver samples the intensity of the diffraction pattern at some point between the center of the pattern and the arbitrary limit set by θ_T. An average value of the received power, assuming a uniform probability distribution of the pointing angle, ϵ, may be obtained by considering the spatial power density within the transmitter beamwidth, θ_T, to be constant and equal to the average value of the far-field pattern between half intensity points. From Equation 1-4, for $\alpha' = \theta_T/2 = \lambda_c/2d_T$ the fractional part of the total power, P_A, in the cone defined by θ_T is

$$\left[1 - J_0^2\!\left(\frac{\pi}{2}\right) - J_1^2\!\left(\frac{\pi}{2}\right)\right] = 0.455 \tag{1-9}$$

As shown in Figure 1-6, the transmitted signal power gathered by the receiver optical antenna is equal to the area of the receiver, $\pi d_R^2/4$, multiplied by the spatial power density at the receiver. Thus,

$$(P_R)_{\mathrm{avg}} = (0.455)\tau_a P_A\!\left(\frac{\pi d_R^2/4}{\pi \theta_T^2 R^2/4}\right) = 0.455\,\frac{\tau_a P_A d_T^2 d_R^2}{R^2 \lambda_c^2} \tag{1-10}$$

This represents a pessimistic result since pointing angle error distributions tend to be Gaussian rather than uniform. For the design of laser communication systems the received power is usually conservatively assumed to be given by Equation 1-8 with the expectation that the actual received power will be 70% to 100% larger, thereby giving a comfortable safety margin in the design.

Two types of receiver optical antennas are shown in Figure 1-7. The focusing type of antenna gathers in the laser signal energy and focuses it to a spot on the photodetector surface; the collimating antenna produces a parallel output beam which is designed to be somewhat smaller than the photodetector. In the focusing antenna a photodetector of diameter d_P is placed at the focal point of the lens and establishes a receiver field of view, θ_R, given by

$$\theta_R = \frac{d_P - d_D}{F} \tag{1-11}$$

where F is the lens focal length and d_D is the diameter of the focused incident beam. If the receiver antenna is diffraction limited, the spot size, as defined by the Airy disk, is

$$d_D = \frac{2.44 F \lambda_c}{d_R} \tag{1-12}$$

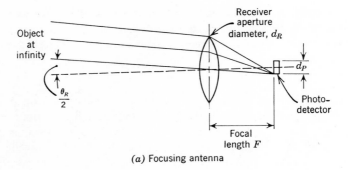

(a) Focusing antenna

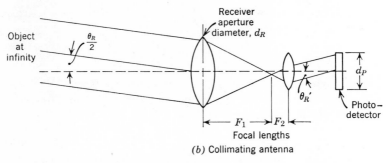

(b) Collimating antenna

Figure 1-7 *Functional diagrams of receiver optical antennas*

If the diffraction limited spot size is much smaller than the photodetector surface diameter, then

$$\theta_R = 2.44\lambda_c \frac{d_P}{d_D d_R} \tag{1-13}$$

The receiver field of view of a collimating antenna, θ_R, is related to the field of view of the photodetector assembly, θ_R', by the expression

$$\theta_R = \frac{F_2}{F_1} \theta_R' \tag{1-14}$$

where F_1 and F_2 are the focal lengths of the collimating lenses [1-1].

The carrier power at the detector surface is reduced by losses in the receiver antenna. These losses, which are described by the receiver transmissivity, τ_r, include attenuation and beam spillover in the antenna. The total signal power at the detector surface, P_C, is then

$$P_C = \tau_r P_R \tag{1-15}$$

Gathering the expressions for beam losses together yields the carrier range relationship between detector signal power and laser power for worst case transmitter beam-pointing conditions.

$$P_C = \frac{\pi^2 \tau_t \tau_a \tau_r d_T^2 d_R^2 P_L}{32 R^2 \lambda_c^2}$$

(1-16)

1.2 COMMUNICATION SYSTEM STATISTICAL MODEL

A generalized statistical model for analysis of a communication system is shown in Figure 1-8. In this model it is assumed that the information source generates a sequence of discrete symbols taken from a finite set of such symbols; analog signals can be put into discrete form by sampling and quantizing. The input signals must then be put into a format suitable for transmission over the channel by the combined operations of coding and communication signal generation. Multiplicative disturbances, consisting of signal attenuation and random phase delays, affect the signal as it passes through the channel. The signals at the output of the channel are detected and decoded to complete the communication link. Background radiation from reflected sunlight, stars, planets, and other sources enters the detector to create external noise which combines with the internal detector noise caused by random emissions.

The system performance of analog links is expressed in terms of the fidelity of the source input signal compared to the destination output signal as measured by mean square error, peak error, or some other suitable criterion. Time quantized pulse and digital systems are usually rated in terms of the probability of detection error of a data sample or bit. To specify measures of detection fidelity, it is necessary to characterize the optical detection process and, specifically, to determine the probability distribution of the output of a photodetector.

Before considering the statistical character of detector emissions, it will be useful to discuss the concept of coherence of light waves [1-2 to 1-4].

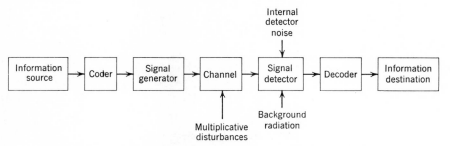

Figure 1-8   *Communication system statistical model*

Coherence is a measure of the ability of a wave to interfere with itself or another wave. Figure 1-9 illustrates two classic experiments which demonstrate the self-coherence of a wave.

In Figure 1-9a, a light beam is split into two components by a semitransparent mirror. These components are totally reflected by mirrors M_1 and M_2 and then recombined at the beam splitter into a single beam which strikes a screen. If M_1 and M_2 are at different distances, S_1 and S_2, from the beam splitter, the resultant beam striking the screen will be composed of the original beam summed with a time-shifted version of itself. Constructive or destructive interference will be evident on the surface of the screen if the original beam is a monochromatic wave with constant phase. If the original beam undergoes phase changes at time instants separated by less than the time interval $(S_1 - S_2)/c$, where c is the velocity of light, the degree of interference will be reduced appreciably. The mean time between phase changes of the original

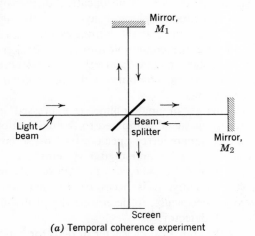

(a) Temporal coherence experiment

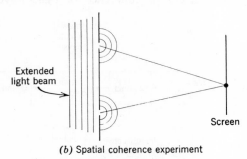

(b) Spatial coherence experiment

Figure 1-9 *Coherence experiments*

beam is called the coherence time, τ_c. The coherence time is related to the bandwidth of the light wave, Δf, measured at the half intensity points by

$$\tau_c \equiv \frac{1}{\Delta f} \tag{1-17}$$

A coherence length is defined as $c\tau_c$.

In Figure 1-9b, a beam of extended diameter strikes a plate with narrow slits at points P_1 and P_2. By Huygen's principle, secondary wavelets with phases equal to the incident phase angles emanate from P_1 and P_2. These secondary waves interfere on the surface of the screen. If the medium through which the incident beam passes is inhomogeneous, the wave fronts at P_1 and P_2 will differ to some degree and the interference on the screen will be reduced. In general, as P_1 and P_2 are separated farther, the amount of interference decreases. The distance between P_1 and P_2 for which the degree of interference drops to some specified level is a measure of the spatial coherence of the wave. A coherence area is often defined as the area of the circle formed with points P_1 and P_2 at opposite ends of its diameter.

To determine the emission statistics of a photodetector, let $\mathscr{I}(t)$ be the instantaneous intensity, measured in watts, of quasimonochromatic† light centered at frequency f_c which is incident upon the photodetector. The optical wave may be considered to be composed of discrete energy packets called photons, each of energy hf_c where h is Planck's constant. Thus the radiation consists of $\mathscr{I}(t)/hf_c$ photon arrivals per second. In the photodetector the arrival of a photon results in the emission of an electron, the creation of a charge carrier, or some other measurable electrical event. The probability of detecting a single photoelectron in an infinitesimal time period, dt, is $g\mathscr{I}(t)\,dt$ where g is a proportionality constant dependent upon the photodetection mechanism [1-5].‡

Mandel has shown that the probability of obtaining k photoelectron counts in the finite time interval t to $t + \tau$ due to a general radiative source is the Poisson distribution

$$P(U_{R,\tau} = k; t) = \frac{1}{k!}\left[\int_t^{t+\tau} g\mathscr{I}(t')\,dt'\right]^k \exp\left[-\int_t^{t+\tau} g\mathscr{I}(t')\,dt'\right] \tag{1-18}$$

where $U_{R,\tau}$ is the random variable that represents the number of photoelectron counts [1-6]. The time interval, τ, is related to the bandwidth of the electrical filter following the photodetector.

The average number of photoelectron counts in a detection period τ due to any source of optical radiation is the time average over τ of the ensemble

---

† Quasimonochromatic light is narrow frequency band optical radiation for which the bandwidth at the half intensity points is much smaller than the mean frequency.

‡ For most photodetectors, $g = \eta/hf_c$ where η is the detector quantum efficiency.

average of $P(U_{R,\tau} = k; t)$,†

$$\mu_{R,\tau} = \sum_{k=0}^{\infty} kP(U_{R,\tau} = k; t) \tag{1-19}$$

From Appendix C the mean of a Poisson distribution gives

$$\mu_{R,\tau} = g \int_{t}^{t+\tau} \mathscr{I}(t') \, dt' = \alpha \widetilde{\mathscr{I}} \tau \tag{1-20}$$

where $\widetilde{\mathscr{I}}$ is the time-averaged optical wave intensity over τ.

The probability distribution $P(U_{R,\tau} = k; t)$ is a random function of time since $\mathscr{I}(t)$ is a random function. The required stationary counting distribution $P(U_{R,\tau} = k)$, which characterizes the detection process, is found by taking the time average or statistical average of $P(U_{R,\tau} = k; t)$. If a statistical average is taken the photoelectron counting distribution becomes

$$P(U_{R,\tau} = k) = \frac{1}{k!} \int_{0}^{\infty} \left[\int_{t}^{t+\tau} g\mathscr{I}(t') \, dt' \right]^{k} \exp \left[-\int_{t}^{t+\tau} g\mathscr{I}(t') \, dt' \right] P(\mathscr{I}) \, d\mathscr{I} \tag{1-21}$$

where $P(\mathscr{I})$ is the probability distribution of the intensity of the optical wave incident upon the detector. In general, $P(U_{R,\tau} = k)$ is not a Poisson distribution.

Before the general form of $P(U_{R,\tau} = k)$ is determined, two limiting cases of great interest will be considered. These are the situations in which the coherence time, τ_c, of the optical radiation is much greater than the integration period ($\tau_c \gg \tau$) and the opposite case ($\tau \gg \tau_c$). The former case is characteristic of laser radiation while the latter applies to incoherent background radiation.

Consider first the situation in which background radiation, made quasi-monochromatic by an optical filter, is incident upon a photodetector. Background radiation, arising from thermal sources, may be considered to be composed of a multitude of waves added together in random phase. By the central limit theorem of statistics, the wave amplitude of the radiation is found to be a Gaussian random variable [1-6, 1-7]. Then by a transformation of variables, the probability distribution of the instantaneous intensity is the exponential distribution

$$P(\mathscr{I}_B) = \frac{1}{\widetilde{\mathscr{I}}_B} \exp -\left(\frac{\mathscr{I}_B}{\widetilde{\mathscr{I}}_B} \right) \tag{1-22}$$

---

† The notation $[\widetilde{X(t)}]$ represents the time average of a function,

$$[\widetilde{X(t)}] = \frac{1}{\tau} \int_{-\tau/2}^{\tau/2} [X(t)] \, dt$$

where $\widetilde{\mathscr{I}}_B$ is the average background radiation intensity over the time period τ. For the case $\tau \gg \tau_c$, the variance of the photoelectron counting distribution is found to be

$$\sigma^2(U_{B,\tau}) = g\widetilde{\mathscr{I}}_B\tau + (g\widetilde{\mathscr{I}}_B\tau)^2 \frac{\tau_c \mathscr{A}_c}{\tau \mathscr{A}} \tag{1-23}$$

where $\mathscr{A}$ is the area of the detector and $\mathscr{A}_c$ is the coherence area [1-8]. The coherence time of background radiation from a thermal source is approximately 10^{-12} sec and the ratio $\mathscr{A}_c/\mathscr{A}$ is usually less than 10^{-3}. Hence, for all practical values of the decision interval τ, the second term of Equation 1-23 is negligible. The resulting variance in photoelectron counts is therefore the same as if the distribution $P(U_{B,\tau} = k)$ were Poisson.

For highly coherent radiation from a laser the condition $\tau \ll \tau_c$ is fulfilled. Consider first a laser that emits at a single frequency—a single-mode laser—and operates far above its threshold of oscillation. The probability distribution of the laser intensity is often assumed to be of the ideal form.

$$P(\mathscr{I}_S) = \delta(\mathscr{I}_S - \widetilde{\mathscr{I}}_S) \tag{1-24}$$

where $\widetilde{\mathscr{I}}_S$ is the average laser intensity over τ. The photoelectron counting distribution of Equation 1-21 then reduces to the stationary Poisson distribution

$$P(U_{S,\tau} = k) = \frac{(g\widetilde{\mathscr{I}}_S\tau)^k \exp\{-(g\widetilde{\mathscr{I}}_S\tau)\}}{k!} \tag{1-25}$$

with mean and variance

$$E[U_{S,\tau}] = \sigma^2(U_{S,\tau}) = g\widetilde{\mathscr{I}}_S\tau \tag{1-26}$$

The variance term $g\widetilde{\mathscr{I}}_S\tau$ is often called shot noise. This noise, as shown in Chapter 8, has a flat frequency spectrum.

Another characterization of the probability density of the intensity of a single-mode laser is a Gaussian distribution [1-9]. This assumption leads to a rather complicated expression for the photoelectron counting distribution in terms of Hermite polynomials. While the Gaussian distribution gives somewhat better agreement with experimental results [1-10] than the delta function distribution, the latter is within the bounds of experimental error. Since the delta function distribution yields a much more mathematically tractable result for the photoelectron counting distribution, Equation 1-24 will be taken as the model for the probability density of a single-mode laser far above threshold.

In a multimode laser the condition $\tau \ll \tau_c$ remains valid. The instantaneous optical wave amplitude is a sum of components from the different laser

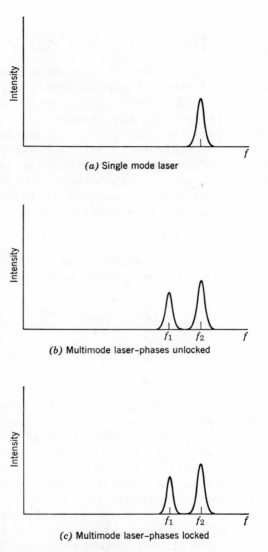

Figure 1-10   Spectra of lasers and photodetector current

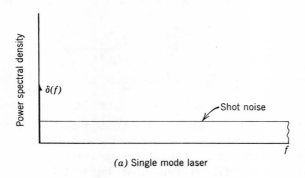

(a) Single mode laser

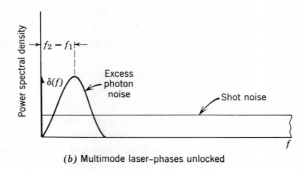

(b) Multimode laser—phases unlocked

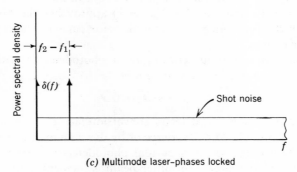

(c) Multimode laser—phases locked

Figure 1-10   (*Continued*)

modes; the amplitude distribution tends to be Gaussian if the modes oscillate independently. This leads to an exponential intensity distribution from which the photoelectron counting distribution is found to be the Bose-Einstein distribution [1-6].

$$P(U_{R,\tau} = k) = \frac{(g\widetilde{\mathscr{I}}\tau)^k}{[1 + g\widetilde{\mathscr{I}}\tau]^{k+1}} \tag{1-27}$$

The variance of the Bose-Einstein distribution

$$\sigma^2(U_{R,\tau}) = (g\widetilde{\mathscr{I}}\tau) + (g\widetilde{\mathscr{I}}\tau)^2 \tag{1-28}$$

is much larger than the variance of the Poisson distribution since $g\widetilde{\mathscr{I}}\tau$ is usually much greater than unity. The second term of Equation 1-28 is often called excess photon noise. The frequency spectrum of the excess photon noise is approximately Gaussian in shape. Figure 1-10 is a sketch of the combined spectra of a laser and the photodetector output current [1-11]. The excess photon noise is eliminated if the laser modes are phase locked and do not oscillate independently.

Returning now to Equation 1-18, when the optical energy incident upon the photodetector is defined as

$$F(\tau, t) = \int_t^{t+\tau} \mathscr{I}(t') \, dt' \tag{1-29}$$

the counting distribution $P(U_{R,\tau} = k)$ may be investigated through its cumulants [1-6]. By this technique it is found that when the average light intensity $\widetilde{\mathscr{I}}$ is small, $P(U_{R,\tau} = k)$ becomes Poisson. This is an important result; in the interest of efficiency, communication systems are generally designed to operate at low received power levels. In the other extreme where $\widetilde{\mathscr{I}}$ is large, the distribution of $P(U_{R,\tau} = k)$ approaches the distribution of $gqF(\tau, t)$ which is the integrated photodetector current where q is the unit of electronic charge.

The instantaneous output current of a photodetector with a general optical radiative source incident upon it is

$$i_P = gq\widetilde{\mathscr{I}} \equiv \mathscr{D}\widetilde{\mathscr{I}} \tag{1-30}$$

where $\mathscr{D}$ is defined to be the detector conversion factor. If the detector input intensity is not modulated at its source or by the channel, Equation 1-30 also represents the average or direct current photodetector current. In the case that an unmodulated laser transmitter beam is incident upon the detector, the average detector output current is

$$I_S = gq\widetilde{\mathscr{I}}_S \equiv \mathscr{D}P_C \tag{1-31}$$

where P_C is the average unmodulated laser carrier power as specified by

Equation 1-16. Equation 1-31 represents the intensity-to-current conversion model of photodetection. The photodetector can also be characterized by $M_{S,\tau}$, the average number of laser carrier photon arrivals per time period τ.

$$M_{S,\tau} = \frac{\tau P_C}{hf_c} \tag{1-32}$$

A group of $M_{S,\tau}$ photons generates an average of $\eta M_{S,\tau}$ photoelectron counts where η, ($\eta \leq 1$), is the photodetector quantum efficiency. Some photons are absorbed without the generation of a photoelectron count due to absorption losses in the photodetector material. The photodetector may thus be described by a photon-to-electron converter model

$$\mu_{S,\tau} = \eta M_{S,\tau} \tag{1-33}$$

where $\mu_{S,\tau}$ is the average number of photoelectron counts obtained by the detector in the time interval τ due to the laser carrier.

REFERENCES

1-1. Born, M. and Wolf, E. *Principles of Optics.* Macmillan, New York, 1964.

1-2. Mandel, L. and Wolf, E. "Coherence Properties of Optical Fields." *Reviews of Modern Physics*, **37** (2), 231–287, Apr. 1965.

1-3. Hodara, H. "The Concept of Coherence and Its Application to Lasers." *Proceedings 3rd Symposium Quantum Electronics*, pp. 121–137, 1963.

1-4. Wolf, E. "Recent Researches on Coherence Properties of Light." *Proceedings 3rd Symposium Quantum Electronics*, pp. 13–34, 1963.

1-5. Mandel, L. "Fluctuations of Photon Beams and their Correlations." *Proceedings Physical Society*, **71**, 1037–1048, 1958.

1-6. Mandel, L. "Fluctuation of Light Beams." *Progress in Optics*. E. Wolf, ed. John Wiley, New York, 1963.

1-7. Mandel, L. "Some Coherence Properties of Non-Gaussian Light." *Proceedings 3rd Symposium Quantum Electronics*, pp. 101–109, 1963.

1-8. Hodara, H. "Statistics of Thermal and Laser Radiation." *Proceedings IEEE*, **53** (7), 696–704, July 1965.

1-9. Bedard, G. "Statistical Properties of Laser Light." *Physics Letters*, **21** (1), 32–33, Apr. 1966.

1-10. Arecchi, F. T. and Berne, A. "High-Order Fluctuations in a Single-Mode Laser Field." *Physical Review Letters*, **16** (1), 32–35, Jan. 1966.

1-11. Hodara, H. and George, N. "Excess Photo Noise in Multimode Lasers." *IEEE Journal of Quantum Electronics*, **QE-2** (9), 337–340, Sept. 1966.

chapter 2

MODULATION
AND DEMODULATION

Modulation is the process of varying the amplitude, intensity, frequency, phase, or polarization of a carrier to convey an information signal. Demodulation is the inverse process, that of extracting the information signal from the carrier.

This chapter presents a unified description of the various types of modulation used throughout the electromagnetic spectrum. Some modulation methods are restricted to optical frequency carriers, while other methods are applicable only to radio frequencies. The latter class of methods is of concern in this book because of subcarrier modulation systems. In subcarrier systems an information signal modulates a radio subcarrier which in turn modulates an optical carrier.

Short descriptions of the operating principles of optical and radio receivers are given to provide a background for the subsequent chapters. Chapter 10 contains detailed descriptions of the operation of optical receivers, and references [2-1] to [2-4] provide further information on the operation of radio receivers.

2.1 MODULATION METHODS

Modulation of an optical carrier differs from modulation of a radio frequency carrier primarily because of the characteristics and limitations of the devices used for performing the modulation operation. At optical frequencies, many modulators are designed to operate directly upon the carrier intensity (amplitude squared of the electric field) rather than the amplitude of the carrier as is common with radio frequency modulators. Furthermore, since optical detectors respond to the carrier intensity, continuous amplitude modulation is of limited usefulness because of the inherent nonlinearity between the modulation signal and the detector output. Phase modulators exist for the optical spectrum, but their application is limited by difficulties in demodulation resulting from frequency instability of the carrier transmitter and the receiver local oscillator. Polarization modulation is easily accomplished at optical frequencies, but it has not proven feasible at radio frequencies.

Laser modulation methods are classified in Table 2-1 as analog, pulse, and digital forms of modulation. In analog-modulation systems, an analog information signal, $M(t)$, continuously varies the amplitude, frequency, phase, intensity, or polarization of the carrier. Pulse-modulation systems employ an intermittent carrier in which an electrical parameter of the carrier, the carrier's duration, or its time of occurrence is varied to convey information. In most pulse-modulation systems the information signal is time-sampled and there is a one-to-one correspondence between data samples and carrier pulses. If $M(t)$ is an information signal whose highest frequency component is B_O, from the sampling theorem it is known that $M(t)$ can be reconstructed from its samples, $M(t_n)$, time spaced $1/2B_O$ seconds apart [2-1]. In practice most signal sources are not bandlimited; sampling must be performed at a higher rate to prevent errors in data reconstruction. The amplitude of a data sample is often restricted to a certain set of levels by a quantization operation to permit the use of digital storage and processing equipment. Data quantization is required for all digital forms of modulation. For digital modulation systems, a discrete set of symbols is assigned to code each quantized data sample. Usually the code set consists of two symbols called "ones" and "zeros," and the data coding is called pulse code modulation (PCM).

Type of modulation	Analog	Pulse	Digital
Information signal	Time continuous	Time continuous or sampled	Time sampled
Carrier parameter (amplitude, intensity, frequency, phase, or polarization)	Continuous	Continuous or quantized	Quantized and coded
Example	Intensity modulation	Pulse intensity modulation	PCM intensity modulation

Table 2-1. CLASSIFICATION OF LASER MODULATION METHODS

Figure 2-1 illustrates the classification of intensity modulation systems. In an analog intensity modulation (IM) system (Figure 2-1a), the intensity of the transmitted laser beam is directly proportional to the time-varying, continuous information signal. Quantized time samples of the information signal are shown in Figure 2-1b. For quantized pulse intensity modulation (PIM) the intensity of the carrier is set proportional to the quantized amplitude of a data sample for a specified time duration. Figure 2-1c illustrates a binary code for the quantized data samples. For this particular example the binary number system representation of the data sample amplitudes was chosen for the code. The "ones" and "zeros" of the binary-coded information signal are represented as maximum and zero carrier intensity, respectively.

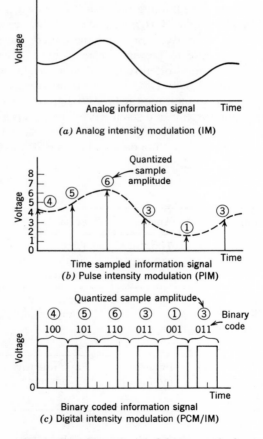

(a) Analog intensity modulation (IM)

(b) Pulse intensity modulation (PIM)

Binary coded information signal
(c) Digital intensity modulation (PCM/IM)

Figure 2-1   *Intensity modulation methods*

The following is a description of the most common methods of laser modulation. More detailed descriptions of the modulation systems are given in the next section.

Analog Methods

AM—analog Amplitude Modulation—carrier electric field amplitude is set proportional to information signal amplitude.

FM—analog Frequency Modulation—carrier instantaneous frequency is set proportional to information signal amplitude.

PM—analog Phase Modulation—carrier phase angle is set proportional to information signal amplitude.

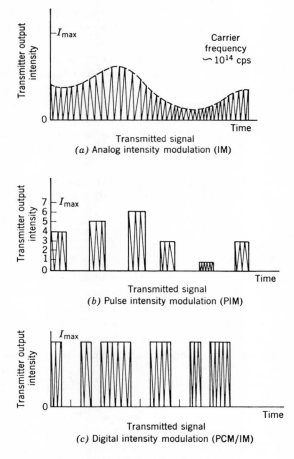

Figure 2-1   (*Continued*)

IM—analog Intensity Modulation—carrier intensity is set proportional to information signal amplitude.

PL—analog Polarization Modulation—linear type: angle of linear carrier polarization with respect to reference axis is set proportional to information signal amplitude; circular type: ratio of carrier intensity in right-to-left polarization states is set proportional to information signal amplitude.

Pulse Methods

PAM—continuous or quantized Pulse Amplitude Modulation—pulsed carrier electric field amplitude is set proportional to information signal sample amplitude.

PFM—continuous or quantized Pulse Frequency Modulation—pulsed carrier frequency is set proportional to information signal sample amplitude.

PIM—continuous or quantized Pulse Intensity Modulation—pulsed carrier intensity is set proportional to information signal sample amplitude.

PDM—continuous or quantized Pulse Duration Modulation—pulsed carrier duration, with respect to start of sample period, is set proportional to information signal sample amplitude.

PPM—continuous or quantized Pulse Position Modulation—time delay of initiation of a short-duration carrier pulse is set proportional to information signal sample amplitude.

PRM—Pulse Rate Modulation—number of short-duration carrier pulses per unit time period is set proportional to information signal amplitude.

Digital Methods

PCM/IM(PCM/AM)—PCM Intensity (amplitude) Modulation, also called PCM/ASK, amplitude shift keying—carrier intensity (amplitude) is set at maximum to represent a "one" bit or at minimum to represent a "zero" bit of binary code of information signal sample amplitude.

PCM/FM—PCM Frequency Modulation, also called PCM/FSK, frequency shift keying—carrier frequency is set at one of two possible values to represent "one" or "zero" bit of binary code of information sample amplitude.

PCM/PM—PCM Phase Modulation, also called PCM/PSK, phase shift keying—carrier phase angle is set at a phase angle of zero or π radians with respect to a phase reference to represent "one" or "zero" bit of binary code of information signal amplitude.

PCM/PL—PCM Polarization Modulation—linear type: carrier is set in vertical polarization to represent "one" bit and horizontal polarization to represent "zero" bit of binary code of information signal sample amplitude; circular type: carrier is set in right circular polarization to represent "one" bit and left circular polarization to represent "zero" bit of binary code of information signal amplitude.

2.2 MODULATION SPECTRA

Amplitude, intensity, frequency, phase, and polarization modulation can be defined in terms of the parameters of the electric field components which are set proportional to a bipolar modulation signal, $M(t)$, that is normalized to unity ($|M(t)| \leq 1$). The following sections describe the electric field equations for an unmodulated carrier and for various types of carrier modulation.

Unmodulated Carrier

The instantaneous electric field, $E(t)$, of an unmodulated carrier may be represented as

$$E(t) = A_c \cos(\omega_c t + \Phi_c) \tag{2-1}$$

where

$$A_c = \text{carrier electric field amplitude}$$

$$\frac{\omega_c}{2\pi} = f_c = \text{carrier frequency}$$

$$\Phi_c = \text{carrier phase angle}$$

For a carrier wave propagating along the Z axis, the electric field components along the X and Y axes are

$$E_X(t) = A_X \cos(\omega_c t + \Phi_X) \tag{2-2}$$

$$E_Y(t) = A_Y \cos(\omega_c t + \Phi_Y) \tag{2-3}$$

where A_X and A_Y are the carrier electric field amplitudes and Φ_X and Φ_Y are the carrier phase angles along the X and Y axes. The instantaneous carrier intensity, $C(t)$, is defined as the square of the electric field

$$C(t) \equiv E^2(t) = A_c^2 \cos^2(\omega_c t + \Phi_c) \tag{2-4}$$

The average carrier power, P_C, is the time average of $C(t)$ over a carrier wave period

$$P_C = \widetilde{C(t)} = \tfrac{1}{2}A_c^2 \tag{2-5}$$

Analog Amplitude Modulation

The carrier electric field amplitude is set proportional to the modulation signal for analog amplitude modulation

$$A_c \sim M(t) \tag{2-6}$$

Conventional double-sideband amplitude modulation results when the carrier electric field, $E(t)$, is multiplied by $\frac{1}{2}[1 + M(t)]$, yielding†

$$E_M(t) = \frac{A_c}{2} [1 + M(t)] \cos \omega_c t \tag{2-7}$$

If $M(t)$ in Equation 2-7 varies over the full range ± 1, the carrier is said to be 100% amplitude modulated. An AM modulation index, M_{AM}, describing the degree of modulation is defined as

$$M_{AM} = \frac{[E_M(t)]_{max} - A_c/2}{A_c/2} \tag{2-8}$$

where $[E_M(t)]_{max}$ is the peak value of the carrier electric field amplitude.

Since the electric field of the modulated carrier is linearly related to the modulation signal, the frequency spectrum of $E_M(t)$ may be investigated by a single sine wave component of $M(t)$. If $M(t)$ is a sine wave of frequency $f_m = \omega_m/2\pi$, the resulting electric field upon expansion is

$$E_M(t) = \frac{A_c}{2} \left[\cos \omega_c t + \frac{1}{2} \cos (\omega_c + \omega_m)t + \frac{1}{2} \cos (\omega_c - \omega_m)t\right] \tag{2-9}$$

Figure 2-2a illustrates the time and frequency representation for sinusoidal amplitude modulation of the electric field. Only the sidebands contain information about the modulation signal. If the carrier electric field is directly multiplied by the modulation signal, $M(t)$, the result, shown in Figure 2-2b, is

$$E_M(t) = A_c M(t) \cos \omega_c t \tag{2-10}$$

and represents double-sideband suppressed carrier amplitude modulation. For sine wave modulation the carrier electric field becomes

$$E_M(t) = \frac{A_c}{2} [\cos (\omega_c + \omega_m)t + \cos (\omega_c - \omega_m)t] \tag{2-11}$$

Since both sidebands carry the same information, one of them may be deleted by filtering to conserve bandwidth, yielding single-sideband suppressed carrier amplitude modulation. With both single- and double-sideband suppressed carrier AM, it is necessary to reinsert the carrier at the receiver

---

† The carrier phase is neglected since it is unaffected by amplitude modulation.

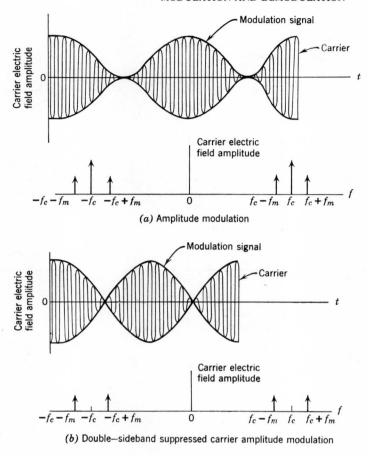

(a) Amplitude modulation

(b) Double–sideband suppressed carrier amplitude modulation

Figure 2-2 *Time and frequency representations of amplitude modulation*

for proper demodulation without distortion. The relative bandwidths required for the three types of amplitude modulation are shown in Figure 2-3.

Analog Frequency Modulation

The instantaneous carrier frequency which is the time derivative of the instantaneous phase angle, is set proportional to the modulation signal in analog frequency modulation

$$\frac{d\Phi_c}{dt} \sim M(t) \qquad (2\text{-}12)$$

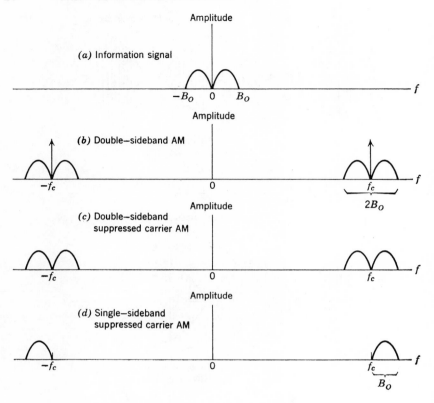

Figure 2-3 *Comparision of AM frequency spectra*

The conventional form for representing frequency modulation is

$$E_M(t) = A_c \cos \left[\omega_c t + \omega_d \int M(t)\, dt \right] \qquad (2\text{-}13)$$

where $\omega_d/2\pi$ is the peak frequency deviation from the carrier frequency, f_c [2-3]. It is difficult to obtain a general expression for the electric field spectrum for an arbitrary modulation signal because of the nonlinear relationship between the modulation signal and the electric field. Spectral representations of a frequency modulated wave are usually given in terms of a single sine wave component of an information signal.

For sinusoidal modulation with $M(t) = \cos \omega_m t$, the frequency modulated electric field is

$$E_M(t) = A_c \cos \left(\omega_c t + \frac{\omega_d}{\omega_m} \sin \omega_m t \right) \qquad (2\text{-}14)$$

The ratio ω_d/ω_m is the ratio of the peak frequency to the modulation frequency and is called the deviation ratio or frequency modulation index, M_{FM}. Expanding $E_M(t)$ by a trigonometric identity yields

$$E_M(t) = A_c \cos \omega_c t \cos [M_{FM} \sin \omega_m t] - A_c \sin \omega_c t \sin [M_{FM} \sin \omega_m t] \qquad (2\text{-}15)$$

By the identities

$$\cos (A \sin \omega t) = J_0(A) + 2 \sum_{n=1}^{\infty} J_{2n}(A) \cos 2n\omega t \qquad (2\text{-}16)$$

and

$$\sin (A \sin \omega t) = 2 \sum_{n=1}^{\infty} J_{2n-1}(A) \sin (2n - 1)\omega t \qquad (2\text{-}17)$$

the electric field equation can be written as a series of Bessel functions of the first kind.

$$E_M(t) = A_c J_0[M_{FM}] \cos \omega_c t$$

$$+ A_c \sum_{n=1}^{\infty} J_{2n}[M_{FM}] \{\cos [\omega_c + 2n\omega_m]t + \cos [\omega_c - 2n\omega_m]t\}$$

$$+ A_c \sum_{n=1}^{\infty} J_{2n-1}[M_{FM}] \{\cos [\omega_c + (2n - 1)\omega_m]t$$

$$- \cos [\omega_c - (2n - 1)\omega_m]t\} \qquad (2\text{-}18)$$

The electric field is thus found to contain a carrier term plus an infinite number of upper and lower sideband frequencies at multiples of the modulation frequency whose amplitudes are dependent upon the Bessel functions of the frequency modulation index. Figure 2-4 illustrates the electric field spectrum for various values of the modulation index.

If the modulation index is small, i.e., $M_{FM} < \pi/2$, only the carrier and first sidebands have significant amplitude. This type of modulation is called narrow-band FM, and the resulting frequency spectrum is similar to that of conventional double-sideband AM. For narrow-band FM the electric field is

$$E_M(t) = A_c \cos \omega_c t + \frac{A_c}{2} (M_{FM}) \cos (\omega_c + \omega_m)t - \frac{A_c}{2} (M_{FM}) \cos (\omega_c - \omega_m)t$$

$$(2\text{-}19)$$

The bandwidth of a sinusoidally modulated FM wave can be defined

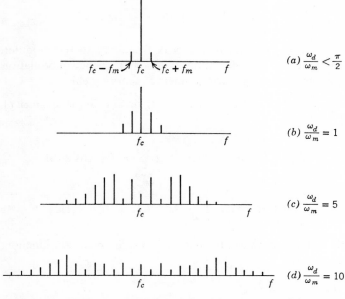

$(a)\ \dfrac{\omega_d}{\omega_m} < \dfrac{\pi}{2}$

$(b)\ \dfrac{\omega_d}{\omega_m} = 1$

$(c)\ \dfrac{\omega_d}{\omega_m} = 5$

$(d)\ \dfrac{\omega_d}{\omega_m} = 10$

Figure 2-4   *Frequency representation of frequency modulation*

as the frequency band for which the sideband amplitudes are below a certain level [2-4]. A rule-of-thumb definition, giving approximately the same results, is that the required bandwidth, B, is

$$B = 2f_m[1 + M_{\mathrm{FM}}] \qquad (2\text{-}20)$$

Phase Modulation

With phase modulation the instantaneous carrier phase, Φ_c, is set proportional to the modulation signal

$$\Phi_c \sim M(t) \qquad (2\text{-}21)$$

The conventional form for phase modulation is

$$E_M(t) = A_c \cos\left[\omega_c t + \Phi_k M(t)\right] \qquad (2\text{-}22)$$

where Φ_k is a phase deviation constant. For sinusoidal modulation the electric field

$$E_M(t) = A_c \cos\left[\omega_c t + \Phi_k \cos \omega_m t\right] \qquad (2\text{-}23)$$

when expanded has the same general form as the frequency modulation spectrum given by Equation 2-18.

Intensity Modulation

For analog intensity modulation the square of the carrier electric field amplitude is set proportional to the modulation signal

$$A_c^2 \sim M(t) \tag{2-24}$$

If the carrier is multiplied by the function $\frac{1}{2}[1 + M(t)]$, the resulting carrier intensity is

$$C_M(t) = \frac{A_c^2}{2} [1 + M(t)] \cos^2 \omega_c t \tag{2-25}$$

An intensity modulation index, M_{IM}, describing the degree of intensity modulation, is defined as

$$M_{IM} \equiv \frac{[C_M(t)]_{max} - (A_c^2/2)}{A_c^2/2} \tag{2-26}$$

where $[C_M(t)]_{max}$ is the peak value of the carrier intensity. For sine wave modulation the carrier intensity, shown in Figure 2-5, is

$$C_M(t) = \frac{A_c^2}{2} [1 + \cos \omega_m t] \cos^2 \omega_c t \tag{2-27}$$

Equation 2-25 represents an ideal expression for intensity modulation. Many intensity modulators produce electric field components of the form (see Section 4.2)

$$E_X(t) = \frac{A_c}{\sqrt{2}} \sin \left[K_{IM} M(t) + \frac{\pi}{4} \right] \cos \omega_c t \tag{2-28}$$

and

$$E_Y(t) = -\frac{A_c}{\sqrt{2}} \sin \left[K_{IM} M(t) + \frac{\pi}{4} \right] \cos \omega_c t \tag{2-29}$$

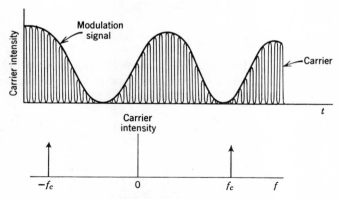

Figure 2-5 *Time and frequency representations of intensity modulation*

where K_{IM} is a physical constant of the modulator. The modulated carrier intensity is then

$$C_M(t) = A_c{}^2 \sin^2 \left[K_{IM} M(t) + \frac{\pi}{4} \right] \cos^2 \omega_c t \qquad (2\text{-}30)$$

and can be written as

$$C_M(t) = \frac{A_c{}^2}{2} \{1 + \sin [2K_{IM} M(t)]\} \cos^2 \omega_c t \qquad (2\text{-}31)$$

For K_{IM} small the expression for the modulated carrier intensity reduces to the ideal form given by Equation 2-25 for less than 100% intensity modulation.

The electric field spectrum for sine wave modulation may be found by letting $M(t) = \sin \omega_m t$ in Equation 2-28 and 2-29. Since the X and Y axis components differ only by a sign, only one of them need be analyzed to determine the carrier frequency spectrum. The X axis, electric field component for sine wave modulation is

$$E_X(t) = \frac{A_c}{2} \{\sin [K_{IM} \sin \omega_m t] + \cos [K_{IM} \sin \omega_m t]\} \cos \omega_c t \quad (2\text{-}32)$$

Expanding in Bessel functions, as for frequency modulation, the electric field equation becomes

$$E_X(t) = \frac{A_c}{2} J_0(K_{IM}) \cos \omega_c t$$

$$+ \frac{A_c}{2} \sum_{n=1}^{\infty} J_{2n}(K_{IM})\{\cos [\omega_c + 2n\omega_m]t + \cos [\omega_c - 2n\omega_m]t\}$$

$$+ \frac{A_c}{2} \sum_{n=1}^{\infty} J_{2n-1}(K_{IM}) \{\sin [\omega_c + (2n - 1)\omega_m]t$$

$$- \sin [\omega_c - (2n - 1)\omega_m]t\} \qquad (2\text{-}33)$$

The frequency spectrum for a sine wave intensity modulated carrier is thus of the same form as a frequency modulated wave. If K_{IM} is small, the X axis electric field is

$$E_X(t) = \frac{A_c}{2} \cos \omega_c t + \frac{K_{IM} A_c}{4} \sin [\omega_c + \omega_m]t - \frac{K_{IM} A_c}{4} \sin [\omega_c - \omega_m]t$$

$$\qquad (2\text{-}34)$$

or

$$E_X(t) = \frac{A_c}{2} [1 + K_{IM} \sin \omega_m t] \cos \omega_c t \qquad (2\text{-}35)$$

Equation 2-35 is equivalent to the electric field representation for sine wave amplitude modulation of a carrier with a modulation signal $M(t) = K_{IM} \sin \omega_m t$. In summary, the frequency spectrum of a sine wave intensity modulated wave is the same as that of an amplitude modulated (or narrow-band FM) wave for a low percentage of intensity modulation; it is equivalent to that of a wide-band frequency modulated wave for a high percentage of intensity modulation.

Polarization Modulation

For linear polarization modulation, in the general field equations, the phase angles along the orthogonal axes are set equal, i.e., $\Phi_X = \Phi_Y = 0$. Then the polarization angle between the orthogonal axis amplitudes is set proportional to the modulating signal

$$K_{PL}M(t) \sim \tan^{-1}\left(\frac{A_Y}{A_X}\right) \tag{2-36}$$

where K_{PL} is a constant of proportionality dependent upon the physical modulation mechanism. The vector sum of the X and Y axis amplitudes is equal to a constant field amplitude

$$A_c \equiv \sqrt{A_X^2 + A_Y^2} \tag{2-37}$$

Combining these two relations yields

$$E_X = A_c \cos\left[K_{PL}M(t)\right] \cos \omega_c t \tag{2-38}$$

$$E_Y = A_c \sin\left[K_{PL}M(t)\right] \cos \omega_c t \tag{2-39}$$

For sine wave modulation the electric field equations are

$$E_X = A_c \cos\left[K_{PL} \cos \omega_m t\right] \cos \omega_c t \tag{2-40}$$

$$E_Y = A_c \sin\left[K_{PL} \cos \omega_m t\right] \cos \omega_c t \tag{2-41}$$

Upon expansion the electric field expressions yield modulation spectral components similar to those of a frequency modulated signal as given in Equation 2-18.

A difficulty with the linear polarization modulation system is that if the transmitter or receiver rotates unpredictably, or if the atmosphere causes random polarization rotations, detection will be impaired. This problem does not exist for the circular polarization modulation system.

In the analog circular polarization modulation system, the carrier energy is placed in carrier components of right and left circular polarization in proportion to the amplitude of the information signal $M(t)$. The intensities of the right and left circular polarization electric field components are ideally

$$C_R(t) = \frac{A_c^2}{2} [1 + M(t)] \cos^2 \omega_c t \qquad (2\text{-}42)$$

and

$$C_L(t) = \frac{A_c^2}{2} [1 - M(t)] \cos^2 \omega_c t \qquad (2\text{-}43)$$

For example, if $M(t) = 1$, the carrier is completely right circularly polarized; if $M(t) = -0.5$, the carrier is 25% right circularly polarized and 75% left circularly polarized. The right and left circularly polarized carrier components may be regarded individually as intensity modulated carriers. Thus the frequency spectrum of a circular polarization modulation carrier is the same as that of an intensity modulation carrier.

Circular polarization modulation is inherently more efficient than intensity modulation for the same peak power transmitted since the carrier power is always transmitted in one polarization state or another. In an intensity modulation system, only one-half of the potential carrier power is transmitted on the average.

Pulse Amplitude Modulation

The carrier electric field representation of a PAM waveform is

$$E_M(t) = \frac{A_c^2}{2} [1 + M(t_n)] \cos \omega_c t \qquad \text{for } t_n \leq t \leq t_n + \tau \qquad (2\text{-}44)$$

where $M(t_n)$ is the amplitude of the information, which may be quantized or continuous, at the sample time t_n and τ is the constant carrier pulse duration.

Pulse Frequency Modulation

The carrier electric field representation of a PFM wave is

$$E_M(t) = A_c \cos \left[\omega_c t + \omega_d \int M(t_n) \, dt \right] \qquad \text{for } t_n \leq t \leq t_n + \tau \qquad (2\text{-}45)$$

Since $M(t_n)$ is constant for a particular data sample, the carrier pulse frequency is simply shifted by some fraction of the maximum frequency deviation for the carrier pulse duration.

Pulse Intensity Modulation

The carrier intensity representation of a PIM wave is

$$C_M(t) = \frac{A_c^2}{2} [1 + M(t_n)] \cos^2 \omega_c t \qquad \text{for } t_n \leq t \leq t_n + \tau \qquad (2\text{-}46)$$

Pulse Duration Modulation

The carrier electric field representation of a PDM wave is

$$E_M(t) = A_c \cos \omega_c t \qquad \text{for } t_n \le t \le t_n + \tau_w \qquad (2\text{-}47)$$

where the time duration τ_w, of the carrier pulse is

$$\tau_w = \frac{\tau_P}{2} [1 + M(t_n)] \qquad (2\text{-}48)$$

The maximum carrier pulse duration is generally restricted to less than the sample period, τ_P, to enable the receiver to detect the carrier pulse leading edge for synchronization purposes. A variant of the PDM system is called pulse width modulation (PWM). In a PWM system, the pulse duration varies symmetrically about the data sample time, t_n.

Pulse Position Modulation

The carrier electric field representation of a PPM wave of duration τ is

$$E_M(t) = A_c \cos \omega_c t \qquad \text{for } t_n + \tau_d \le t \le t_n + \tau_d + \tau \qquad (2\text{-}49)$$

where the time delay, τ_d, of the leading edge of the carrier pulse, with respect to the sample time t_n, is given by

$$\tau_d = \frac{\tau_P}{2} [1 + M(t_n)] \qquad (2\text{-}50)$$

The maximum carrier pulse delay must be less than the sample period, τ_P, in order to prevent pulse overlap in time-adjacent sample periods. In many PPM systems, a marker pulse is transmitted at the beginning of each sample period to provide synchronization. The marker pulse duration is generally wider than the data pulse to provide discrimination between the two.

Pulse Rate Modulation

The PRM system may be either time continuous or time sampled. In the time continuous system the instantaneous rate of generation of carrier pulses is proportional to the information signal amplitude. One method of generating such a waveform would be to frequency modulate a radio frequency (RF) oscillator with the information signal. The frequency of the oscillator and minimum and maximum frequency deviation of the RF modulator would be chosen to match the desired maximum and minimum carrier pulse rates. Then a trigger circuit could be used to generate an optical pulse for every zero crossing of the RF wave.

In a time-sampled PRM system the amplitude of each data sample determines the carrier pulse rate which is maintained over the data sample period. If the data sample amplitude is quantized, the number of carrier pulses per data sample period is made proportional to the sample amplitude. The receiver can then simply "count" carrier pulses to provide demodulation.

PCM Intensity Modulation

The carrier electric field representation of a PCM/IM wave is

$$E_M(t) = A_c \cos \omega_c t \qquad -\frac{\tau_B}{2} \le t \le \frac{\tau_B}{2} \text{ for a "one" bit} \qquad (2\text{-}51)$$

$$E_M(t) = 0 \qquad -\frac{\tau_B}{2} \le t \le \frac{\tau_B}{2} \text{ for a "zero" bit} \qquad (2\text{-}52)$$

where τ_B is the bit period. The frequency spectrum of the PCM/IM wave can be described by the Fourier transform, $g_M(\omega)$, of the modulated electric field [2-5].

$$g_M(\omega) = \frac{A_c \tau_B}{2} \left[\frac{\sin\left[(\omega_c - \omega)(\tau_B/2)\right]}{(\omega_c - \omega)(\tau_B/2)} + \frac{\sin\left[(\omega_c + \omega)(\tau_B/2)\right]}{(\omega_c + \omega)(\tau_B/2)} \right] \qquad (2\text{-}53)$$

The positive frequency component of the frequency spectrum sketched in Figure 2-6 consists of a "sine X over X" function at the carrier frequency. The bandwidth of the wave, B, is generally defined by the frequency interval between the first zero crossings.

$$B = \frac{2}{\tau_B} \qquad (2\text{-}54)$$

PCM Frequency Modulation

For PCM frequency modulation the carrier electric field representation is

$$E_M(t) = A_c \cos \omega_c t \qquad -\frac{\tau_B}{2} \le t \le \frac{\tau_B}{2} \text{ for a "one" bit} \qquad (2\text{-}55)$$

$$E_M(t) = A_c \cos (\omega_c + \Delta\omega)t \qquad -\frac{\tau_B}{2} \le t \le \frac{\tau_B}{2} \text{ for a "zero" bit} \qquad (2\text{-}56)$$

The Fourier spectra of the fields, as shown in Figure 2-7, are the same as the

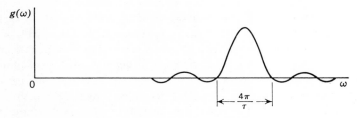

Figure 2-6 *Fourier transform of PCM/IM electric field (positive frequencies only)*

pulse intensity modulation spectrum at the carrier frequency and a shifted version at the carrier frequency plus the frequency deviation $\Delta\omega/2\pi$. The frequency deviation should be several times larger than the highest switching rate, $1/\tau_B$, to prevent significant overlap between the "one" and "zero" spectra.

PCM Phase Modulation

In PCM phase modulation the carrier electric field is

$$E_M(t) = A_c \cos(\omega_c t + \Phi_m) \qquad -\frac{\tau_B}{2} \leq t \leq \frac{\tau_B}{2} \qquad (2\text{-}57)$$

where $\Phi_m = 0$ for a "one" bit and $\Phi_m = \pi$ for a "zero" bit. The Fourier transform of the electric field is [2-5]

$$g_M(\omega) = \frac{A_c \tau_B}{2} \left\{ \frac{\sin[(\omega_c - \omega)(\tau_B/2) + \Phi_m]}{[(\omega_c - \omega)(\tau_B/2) + \Phi_m]} + \frac{\sin[(\omega_c + \omega)(\tau_B/2) + \Phi_m]}{[(\omega_c + \omega)(\tau_B/2) + \Phi_m]} \right\}$$

$$(2\text{-}58)$$

The Fourier spectrum of a PCM/PM wave has the same form as that of a PCM/IM wave, as shown in Figure 2-6.

PCM Polarization Modulation

Bits can be represented as any two orthogonal linear polarization states in the linear PCM/PL modulation system. A common representation is to set the orthogonal electric fields at $\pm 45°$ with respect to a reference coordinate system, usually the crystallographic axes of an electro-optic modulator. For this convention the electric fields for linear PCM/PL are

$$E_X(t) = \frac{A_c}{\sqrt{2}} \cos \omega_c t \qquad -\frac{\tau_B}{2} \leq t \leq \frac{\tau_B}{2} \qquad (2\text{-}59)$$

$$E_Y(t) = \frac{A_c}{\sqrt{2}} \cos(\omega_c t + \Phi_m) \qquad -\frac{\tau_B}{2} \leq t \leq \frac{\tau_B}{2} \qquad (2\text{-}60)$$

where $\Phi_m = 0$ for a "one" bit and $\Phi_m = \pi$ for a "zero" bit.

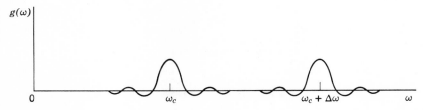

Figure 2-7   *Fourier transform of PCM/FM electric field (positive frequencies only)*

In the circular PCM/PL system, bits are represented as light having right or left circular polarization. The electric field representations are

$$E_X = \frac{A_c}{\sqrt{2}} \cos \left(\omega_c t - \frac{\pi}{4} + \Phi_m \right)$$ (2-61)

$$E_Y = \frac{A_c}{\sqrt{2}} \cos \left(\omega_c t - \frac{\pi}{4} - \Phi_m \right)$$ (2-62)

where $\Phi_m = +\pi/4$ for a "one" bit and $\Phi_m = -\pi/4$ for a "zero" bit. The Fourier spectra of the X and Y axis electric fields for both linear and circular polarization modulation are identical to the general form for a PCM/PM wave, as given by Equation 2-58.

2.3 DEMODULATION METHODS

Optical demodulation operations include transformations that detect amplitude, intensity, frequency, phase, or polarization modulation of the

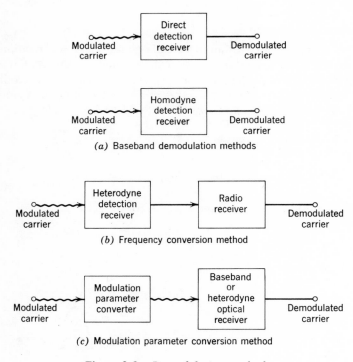

(a) Baseband demodulation methods

(b) Frequency conversion method

(c) Modulation parameter conversion method

Figure 2-8   *Demodulation methods*

carrier and reconstruct the information signal which has modulated the carrier. The three basic types of optical demodulation methods are illustrated by the block diagrams of Figure 2-8. The baseband demodulation systems employ a direct or homodyne detection optical receiver which shifts the modulation spectrum from the carrier frequency to baseband. With the frequency conversion method the information signal spectrum is translated from the optical carrier frequency to a lower radio frequency by an optical heterodyne receiver. A conventional radio receiver then provides demodulation. In the modulation parameter conversion method one type of carrier modulation is converted to another; demodulation is performed on the converted parameter. For example, a frequency modulated optical carrier may be converted to an intensity modulated optical carrier which is demodulated by a baseband or heterodyne optical receiver. The operating principles of the demodulation components illustrated in Figure 2-8 are briefly described in the following sections for noise-free operation.

Direct Detection Optical Receiver

Figure 2-9a is a block diagram of a direct detection optical receiver. In the receiver the laser carrier passes through an optical bandpass filter, which serves to reject background radiation, and impinges on the surface of a photodetector. The photodetector produces an output *current* proportional to the instantaneous *intensity* of the carrier; it may be regarded as a linear intensity-to-current converter or a quadratic (square law) converter of optical electric field-to-detector current. An electrical low pass filter, which has a bandwidth sufficient to pass the information signal, follows the photodetector and limits the amount of photodetector noise.

The photodetector output current, i_P, is proportional to the time average of the instantaneous carrier intensity, $C(t)$, over the carrier period

$$i_P = \mathscr{D}\widetilde{C(t)} = \mathscr{D}A_c{}^2 \widetilde{\cos^2[\omega_c t + \Phi_c]} \qquad (2\text{-}63)$$

where $\mathscr{D}$ is the detector conversion factor which is dependent upon the type of photodetector employed. The time average over the carrier period is independent of the carrier frequency or phase and is equal to one-half. Hence the photodetector current is

$$i_P = \frac{\mathscr{D}A_c{}^2}{2} \qquad (2\text{-}64)$$

Since the photodetector response is insensitive to the frequency, phase, or polarization of the carrier over its operating regime, the direct detection receiver, by itself, is only useful for AM or IM demodulation.

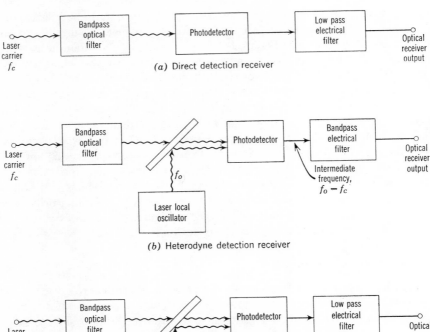

(a) Direct detection receiver

(b) Heterodyne detection receiver

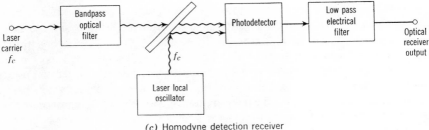

(c) Homodyne detection receiver

Figure 2-9 *Optical receivers*

Heterodyne Detection Optical Receiver

In an optical heterodyne receiver, as shown in Figure 2-9b, a beam-splitting mirror or similar device spatially combines the carrier with an unmodulated laser local oscillator beam whose frequency differs from that of the carrier by the desired intermediate frequency (IF). If the beams are spatially well aligned, optical interference will take place on the photodetector surface. The detector current will then be proportional to the square of the sum of the electric fields of the carrier and the local oscillator. This inherent squaring operation of the photodetector produces a detector current at the intermediate frequency which contains the information signal modulation. The modulation signal on this intermediate frequency (generally in the region of 1 MHz to 1 GHz) is extracted by a conventional radio frequency receiver.

If the carrier and local oscillator beams are aligned perpendicular to the photodetector surface, their electric fields may be written, respectively, as

$$E(t) = A_c \cos [\omega_c t + \Phi_c] \tag{2-65}$$

and

$$L(t) = A_o \cos [\omega_o t + \Phi_o] \tag{2-66}$$

The photodetector output current is then

$$i_P = \mathscr{D}[\overbrace{E(t) + L(t)}]^2 \tag{2-67}$$

which, when expanded, yields

$$i_P = \mathscr{D}\{A_c{}^2 \overbrace{\cos^2 [\omega_c t + \Phi_c]} + A_o{}^2 \overbrace{\cos^2 [\omega_o t + \Phi_o]}$$

$$+ A_c A_o \overbrace{\cos [(\omega_o - \omega_c)t + (\Phi_o - \Phi_c)]}$$

$$+ A_c A_o \overbrace{\cos [(\omega_o + \omega_c)t + (\Phi_o + \Phi_c)]}\} \tag{2-68}$$

In Equation 2-68 the first two terms represent constant or direct current components of the detector output. The amplitude of the difference frequency term may be regarded as constant over the short-duration carrier time average. The bandpass filter centered at the difference frequency then passes an instantaneous current, i_{IF}, equal to

$$i_{\mathrm{IF}} = \mathscr{D}A_c A_o \cos [(\omega_o - \omega_c)t + (\Phi_o - \Phi_c)] \tag{2-69}$$

The IF signal current is thus dependent upon the carrier amplitude, frequency, and phase. Each of these carrier parameters may, therefore, be demodulated by a following radio receiver.

Homodyne Detection Optical Receiver

A homodyne receiver, as shown in Figure 2-9c, employs optical mixing between the carrier and a local oscillator set at the same frequency as the carrier and phase-locked to it. The output of the photodetector is the demodulated signal frequency spectrum shifted to baseband. When the local oscillator frequency is set at the carrier frequency, the photodetector current becomes

$$i_P = \mathscr{D}\left\{\frac{A_c{}^2}{2} + \frac{A_o{}^2}{2} + A_c A_o \cos (\Phi_o - \Phi_c) + A_c A_o \overbrace{\cos [2\omega_c t + (\Phi_o + \Phi_c)]}\right\} \tag{2-70}$$

If the direct current components are neglected, the low pass filter passes a signal current equal to

$$i_S = \mathscr{D}A_c A_o \cos (\Phi_o - \Phi_c) \tag{2-71}$$

Setting the carrier and local oscillator phase angles equal maximizes the signal current for amplitude modulation of the carrier. For phase modulation Φ_c is time varying and if $\Phi_o = \pi/2$ radians, the output signal is proportional to Φ_c for $\Phi_c \ll \pi/2$.

Heterodyne Detection Radio Receivers

Radio heterodyne detection receivers (Figure 2-10) usually employ either a multiplier or square law device to produce an IF sum or difference signal between the carrier and local oscillator. In a product system, shown in Figure 2-10a, the multiplier output, v_H, is

$$v_H = k_R E(t)L(t) = \frac{k_R A_c A_o}{2} \cos\left[(\omega_o - \omega_c)t + (\Phi_o - \Phi_c)\right]$$

$$+ \frac{k_R A_c A_o}{2} \cos\left[(\omega_o + \omega_c)t + (\Phi_o + \Phi_c)\right] \quad (2\text{-}72)$$

where k_R is a constant conversion factor. A bandpass filter following the multiplier selects either the sum or difference IF signal. In most instances, heterodyne conversion is to the lower frequency.

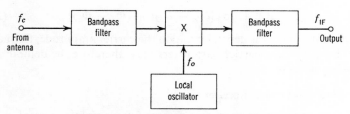

(a) Radio product heterodyne detection receiver

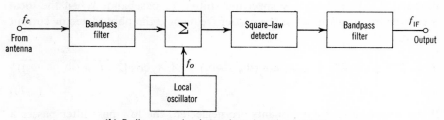

(b) Radio square–law heterodyne detection receiver

Figure 2-10   *Radio heterodyne detection receivers*

In the square-law heterodyne detection receiver of Figure 2-10*b* the carrier and local oscillator waves are summed and then squared. The square-law detector output is then

$$v_H = k_R[E(t) + L(t)]^2 = k_R\{A_c^2 \cos^2[\omega_c t + \Phi_c] + A_o^2 \cos^2[\omega_o t + \Phi_o]$$
$$+ A_c A_o \cos[(\omega_o - \omega_c)t + (\Phi_o - \Phi_c)]$$
$$+ A_c A_o \cos[(\omega_o + \omega_c)t + (\Phi_o + \Phi_c)]\}$$

$$(2\text{-}73)$$

The difference frequency can then be selected by the IF filter following the square-law device.

Baseband Detection Radio Receivers

Four common types of baseband radio receivers are shown in Figure 2-11. Each of these receivers shifts the information signal spectrum modulating the carrier to the baseband frequency.

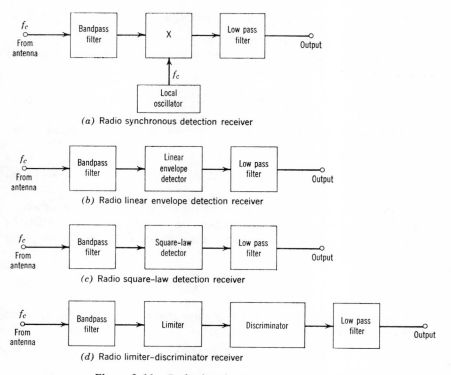

(*a*) Radio synchronous detection receiver

(*b*) Radio linear envelope detection receiver

(*c*) Radio square-law detection receiver

(*d*) Radio limiter–discriminator receiver

Figure 2-11   *Radio baseband detection receivers*

The radio synchronous detection receiver shown in Figure 2-11a operates in the same manner as the radio product heterodyne receiver except that the local oscillator wave is at the same frequency as the carrier. From Equation 2-72 the multiplier output, v_B, is

$$v_B = \frac{k_R A_c A_o}{2} \cos [\Phi_o - \Phi_c] + \frac{k_R A_c A_o}{2} \cos [2\omega_c t + (\Phi_o + \Phi_c)] \quad (2\text{-}74)$$

The low pass filter only passes the first term of Equation 2-74, so the receiver output signal is

$$v_S = \frac{k_R A_c A_o}{2} \cos [\Phi_o - \Phi_c] \quad (2\text{-}75)$$

The local oscillator phase, Φ_o, is set to the carrier phase, Φ_c, for amplitude modulation, and for phase modulation, Φ_o is set to $\Phi_o = \pi/2$.

A linear envelope detection receiver (Figure 2-11b) is designed to produce an output signal proportional to the envelope of the bandpass filter output. The receiver output signal

$$v_S = k_R |E(t)| = k_R A_c \quad (2\text{-}76)$$

is independent of the carrier frequency or phase.

In a radio square-law detection receiver, shown in Figure 2-11c, the bandpass filter output is squared to yield

$$v_B = k_R E^2(t) = k_R \frac{A_c^2}{2} + \frac{k_R A_c^2}{2} \cos [2\omega_c t + 2\Phi_c] \quad (2\text{-}77)$$

The low pass filter following the detector removes the double frequency term, leaving

$$v_S = \frac{k_R A_c^2}{2} \quad (2\text{-}78)$$

The output signal is thus proportional to the instantaneous carrier power.

A radio limiter-discriminator produces an output signal which in the absence of noise is proportional to the time derivative of the carrier phase angle.

$$v_S = k_R \frac{d\Phi_c}{dt} \quad (2\text{-}79)$$

For frequency modulation with an angular frequency deviation, ω_d, from Equation 2-13, the receiver output signal is

$$v_S = k_R \omega_d M(t) \quad (2\text{-}80)$$

where $M(t)$ is the information signal frequency modulating the carrier.

Modulation Parameter Converters

At optical frequencies it is possible to convert a frequency modulated carrier to an intensity modulated carrier with an FM-to-IM converter. Examples of two types of FM-to-IM converters are shown in Figure 2-12. In the skirt system an optical filter with an attenuation characteristic, as shown in Figure 2-12a, is placed before an optical receiver. The filter output intensity is linearly proportional to the carrier frequency over the frequency deviation of the frequency modulated carrier.

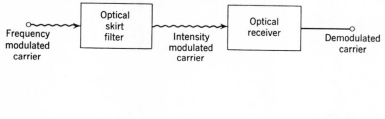

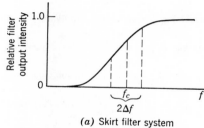

(a) Skirt filter system

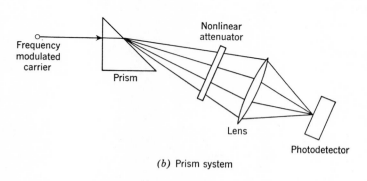

(b) Prism system

Figure 2-12 *FM-to-IM converters*

The prism system employs an ordinary triangular prism which deviates the carrier beam by an angle proportional to the carrier frequency due to the dispersion properties of the prism. The derivative of the deviation angle, δ, with respect to wavelength is inversely proportional to the cube root of the wavelength [2-6].

$$\frac{d\delta}{d\lambda} \sim \lambda^{-3} \tag{2-81}$$

An optical attenuator whose transmissivity is a function of its linear dimension provides attenuation of the carrier beam proportional to its frequency. A lens then gathers the carrier beam energy and focuses it to the surface of a photodetector.

REFERENCES

2-1. Schwartz, M. *Information Transmission, Modulation, and Noise.* McGraw-Hill, New York, 1959.

2-2. Panter, P. F. *Modulation, Noise, and Spectral Analysis,* McGraw-Hill, New York, 1965.

2-3. Schwartz, M., Bennett, W. R., and Stein, S. *Communication Systems and Techniques.* McGraw-Hill, New York, 1966.

2-4. Downing, J. J. *Modulation Systems and Noise.* Prentice-Hall, Englewood Cliffs, N.J., 1964.

2-5. Cuccia, C. L. *Harmonics, Sidebands, and Transients in Communication Engineering.* McGraw-Hill, New York, 1952.

2-6. Morgan, J. *Introduction to Geometrical and Physical Optics.* McGraw-Hill, New York, 1953.

chapter 3

OPTICAL COMPONENTS

This chapter presents a capsule description of the operation of lasers, optical antennas, and other optical components found in laser communication systems. References [3-1] to [3-8] contain greater detail on the theory of operation of these devices and practical problems associated with their construction.

3.1 LASERS

A laser, strictly speaking, is an amplifier of light. When the laser is suitably excited by optical or electrical energy, light of the proper frequency entering the laser cavity is amplified in such a manner that the laser output wave is in phase with the input. Lasers are seldom used as amplifiers since their internal noise (called spontaneous emission) is relatively high. Also there are practical difficulties in making the laser gain high while avoiding oscillation due to inherent energy feedback. The practical utility of a laser is as an oscillator— a generator of coherent light.

The basic principle of laser emission can be described by an energy level diagram of the laser material. Figure 3-1 illustrates such a diagram for a material with three energy levels. Excitation energy is applied to the laser material, causing atoms in the ground state (i.e., atoms whose electrons are

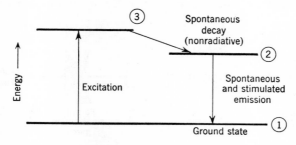

Figure 3-1   *Three energy level diagram of laser material*

45

in orbits near the nucleus) to go to energy level 3. Atoms then very rapidly decay spontaneously to energy level 2 without radiating energy. The energy difference between levels 2 and 3 is dissipated as heat by lattice vibrations within the laser material. From energy level 2, atoms decay to the ground state by spontaneous emission and radiate energy at a frequency, f, given by

$$f = \frac{E_2 - E_1}{h} \tag{3-1}$$

where E_i is the energy of i-th level and h is Planck's constant. If the energy level difference $E_2 - E_1$ is large enough, the radiated energy will be at an optical frequency. Spontaneous emissions are random in time from atom to atom of the material; hence the radiation is not coherent. If the excitation of the material is at a sufficiently high rate, the number of atoms in energy level 2 can be made greater than the number in the ground state. In this instance a "population inversion" is said to exist. Under such conditions the absorption of the material becomes negative (i.e., amplification), and spontaneous emissions stimulate other emissions in time phase. The result is the generation of relatively narrow, spectral-band coherent radiation. For highly stable lasers, emission linewidths of about 10 Hz to 100 Hz have been measured [3-1].

Feedback for oscillation of a laser is achieved by placing the laser in an optical cavity constructed by placing reflecting surfaces at the ends of the laser material (Figure 3-2). One mirror is made partially transmitting to allow useful radiation to escape the cavity. Emissions along the laser axis are reinforced by amplification action on each pass of the laser material. Off-axis emissions are of low energy since they are not multiply amplified.

The emission frequency of a laser is determined by several factors. Oscillation can only occur at frequencies within a spectral interval for which the cavity gain is sufficient to overcome the absorption losses. The center frequency of this spectral interval is set by the mean energy level difference of the stimulated emission transition. The spectral interval of possible oscillation frequencies is broadened by frequency variations due to thermal motion of atoms of the material, called Doppler broadening, and by atomic collisions.

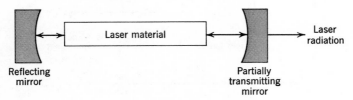

Figure 3-2   Laser cavity

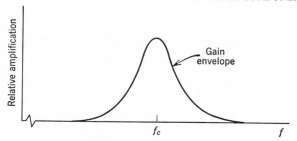

Figure 3-3   *Relative amplification of a laser material*

Figure 3-3 illustrates the relative amplification or gain envelope of a laser material, under broadening, as a function of frequency. The width of the gain envelope at its half amplitude points is typically of the order of 1 GHz.

Resonance conditions of the cavity determine the exact frequencies or modes of oscillation within the gain envelope. Along the axis of the cavity the only frequencies that can exist are those for which the distance between mirrors is an integral number of half wavelengths. For axial modes the resonance condition is

$$\frac{n\lambda_c}{2} = L_c \tag{3-2}$$

where L_c is the distance between mirrors (often the length of the laser material), λ_c is the transmission wavelength, and n is an integer. Thus the possible axial-mode emission frequencies are

$$f_R = \frac{nc}{2L_c} \tag{3-3}$$

where c is the velocity of light. The axial modes are separated by a frequency interval

$$\Delta F = \frac{c}{2L_c} \tag{3-4}$$

For example, a 1-meter laser cavity will support axial modes spaced by 150 MHz. Figure 3-4 shows the axial modes of a laser.

In many applications, multiple axial modes are not desirable. For example, in a heterodyne system the modes mix with each other as well as with the local oscillator, resulting in a loss of useful power. Single-mode operation can be achieved simply by reducing the cavity length so that the low-amplitude modes are below the oscillation threshold. However, with many types of lasers, this results in a reduction of laser power. Techniques have been developed for frequency modulating a laser so that the laser energy is transferred to a single supermode [3-9].

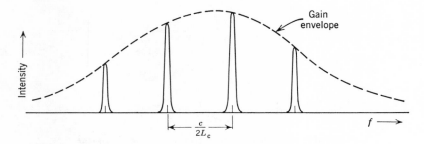

Figure 3-4 *Laser axial modes*

A laser may also oscillate in modes transverse to the axis. Several low-order transverse modes of oscillation are shown in Figure 3-5. Arrows indicate the electric field direction. The designation $TEM_{m,n}$ represents the number of transverse electromagnetic modes as indicated by the examples. The transverse mode structure of a laser is dependent upon the type of cavity mirrors and their alignment [3-10]. With proper design and adjustment the TEM_{00} mode can be selected. Operation in this mode results in a bell-shaped intensity cross section of the beam. With other mode structures the nonuniform illumination of the transmitter aperture results in a far-field beam pattern farther from the diffraction limit.

There are three basic types of lasers: the gas laser, the semiconductor laser, and the solid state laser. Review articles on the theory of operation of these lasers and characteristics of laser materials are to be found in references [3-11] to [3-15]. The following paragraphs briefly describe the operating principles of each type of laser and summarize their advantages and disadvantages for communication purposes.

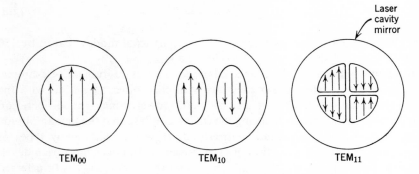

Figure 3-5 *Low-order transverse modes of a laser*

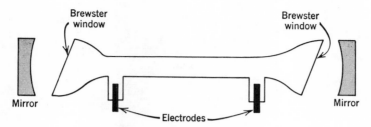

Figure 3-6   *Principal elements of a gas laser*

In a gas laser (Figure 3-6) a gas such as argon, helium-neon, or carbon dioxide, contained in a glass tube, is bombarded by an electron discharge which causes excitation of the atoms. Most gas lasers employ a direct current discharge, though operation with an alternating current discharge is possible. The end windows of the glass tube are usually cut at Brewster's angle ($\approx 57°$) to linearly polarize the output beam. There are two major types of electron excitation: direct electron and atomic collision excitation. In the former, electrons from the discharge strike atoms, causing the upper atomic levels to be populated. Atomic collision provides excitation for lasers containing a mixture of gases, e.g., helium and neon. One of the gases is excited by direct electron collision. Excited atoms of this gas then strike unexcited atoms of the other gas having similar energy levels and provide a population inversion in the second gas.

Figure 3-7 shows the energy level transitions of a helium-neon laser. Helium atoms are excited by electron impact. The helium energy states, labeled 2^3S and 2^1S, are metastable states. They have no allowable transitions to lower levels. When a helium atom in a metastable state collides with an unexcited neon atom in the 3s or 2s state, there is a high probability of energy transfer. In this manner the upper neon energy levels are populated. Transitions from 3s and 2s states to lower energy neon states produce laser oscillation at the wavelengths indicated.

For communications the gas laser offers many desirable properties. Gas lasers exist at many wavelengths throughout the optical spectrum; they are capable of continuous operation without severe heating problems and operate without cooling. Gas lasers are highly stable, possess narrow linewidths, and are highly coherent. The laser beam is near the diffraction limit and, in addition, substantial power is available at some frequencies. The disadvantages of gas lasers are that at some wavelengths the power output and power efficiency are low. Also, to achieve high power, physically long lasers are required.

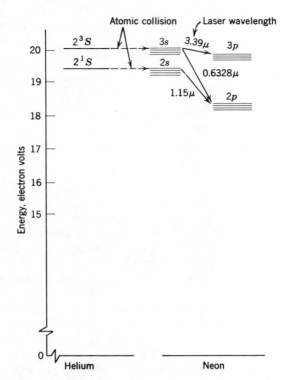

Figure 3-7   *Simplified energy level diagram of a helium-neon laser*

Figure 3-8 contains a sketch of a semiconductor laser. A relatively large current of the order of 10^4 amps/cm² is passed through the junction to provide excitation. The laser emission is perpendicular to the junction with the semiconductor faces forming the cavity.

In semiconductor lasers the energy levels over which stimulated emission occurs are the energy levels of electrons in the conduction and valence bands of the semiconductor materials. A population inversion is obtained by injecting electrons across the junction from the *n*-doped region to the *p*-region. Lasing has been obtained at the infrared frequencies with gallium arsenide and similar semiconductor materials.

The rectangular shape of the junction produces a fan-shaped beam which is not easily collimated. Another disadvantage is that high power and high efficiency usually require cooling of the device. Tens of amperes of driving current are required for laser action; hence, junction heating often limits the use of the laser to pulse operation. The major advantages of the semiconductor

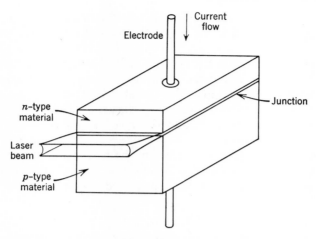

Figure 3-8   Principal elements of a semiconductor laser

laser are its small size and low weight. For an intensity modulation communication system using incoherent detection, semiconductor lasers may be mounted in an array to achieve sizable amounts of transmitter power.

The most widely used solid state laser is the ruby laser. A sketch of the major elements of a ruby laser is contained in Figure 3-9. A ruby rod several centimeters long is encircled by a quartz flashlamp which provides optical excitation. A reflective coating is applied to the ends of the rod to form a cavity. Operation is usually limited by heating of the flashlamp to pulse operation at relatively low repetition rates. Unless rather elaborate feedback mechanisms are employed, the laser output intensity contains spikes. It is possible to place an optical switch in the cavity which introduces enough losses to prevent lasing even though a population inversion is being achieved.

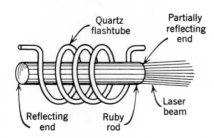

Figure 3-9   Principal elements of a ruby laser

When the optical switch is made transparent the stored energy is released in one giant pulse [3-16]. Such operation is called Q switching and results in megawatt pulses.

Solid-state lasers made of neodymium and other rare earth elements have been developed. Only a few are particularly attractive for communication use due to their low repetition rate capabilities and low efficiencies. An approach to eliminate these difficulties which has met with some success is the pumping of a neodymium-doped crystal with a semiconductor laser rather than a flashlamp [3-17].

In summary, the gas laser is presently the best available source of continuous, highly coherent optical radiation of reasonably high power. Semiconductor lasers are often preferable for pulsed operation.

3.2 OPTICAL ANTENNAS [3-18, 3-19]

The output of a laser transmitter is often collimated by a transmitter optical antenna to reduce the beam divergence. Beam divergence, by the principle of diffraction, is inversely proportional to the antenna aperture diameter.

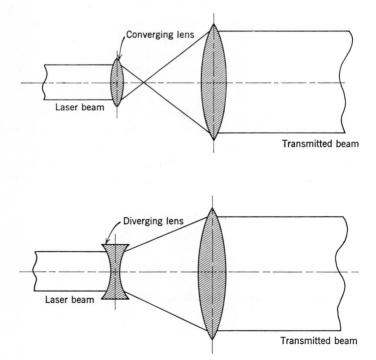

Figure 3-10   Refractive optics transmitter beam collimators

Hence, the larger the antenna, the smaller is the transmitter beamwidth. The transmitter antenna should always be designed as close to the diffraction limit as possible since this results in the smallest size antenna. The narrowness of the beamwidth is limited by beam-pointing considerations to about 1 microradian. At the long wavelength end of the optical spectrum, this establishes a limit of about 10 meters for the transmitter aperture diameter; at the short wavelength end of the spectrum the maximum aperture diameter is about 50 cm. For small size apertures, lenses are practical for forming the transmitter antenna. Two possible configurations are shown in Figure 3-10. Above several centimeters in diameter, cost and weight considerations dictate the use of reflecting optics. Figure 3-11 illustrates two principal configurations.

A receiver antenna can be a reflective or refractive-type collimator that gathers in optical signal energy over an input aperture of diameter d_R and

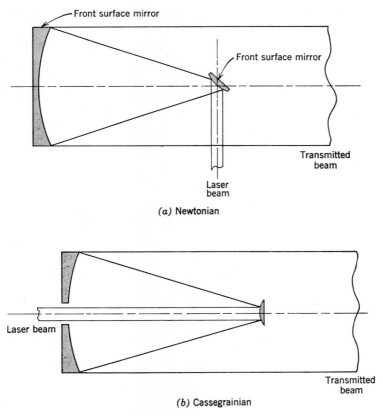

(a) Newtonian

(b) Cassegrainian

Figure 3-11   *Reflective optics transmitter beam collimators*

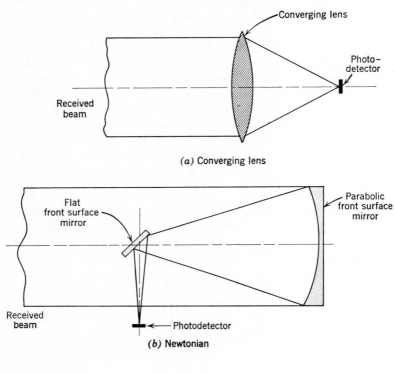

(a) Converging lens

(b) Newtonian

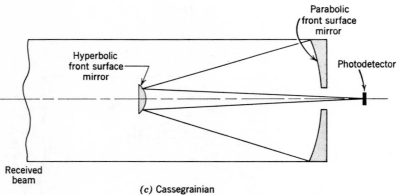

(c) Cassegrainian

Figure 3-12   *Receiver optical antennas*

produces a collimated beam of diameter somewhat smaller than the photo-detector to allow for pointing inaccuracy of the receiver. It is not necessary to produce a collimated beam as an input to the detector; hence a photo-detector may simply be inserted at the focal point of a reflective or refractive receiver antenna of the type shown in Figure 3-12.

In a direct detection optical receiver system, the diameter of the receiver antenna should be as large as possible to gather the maximum amount of signal energy. Since phase is not of importance in direct detection, aberrations of the optical system such as distortion and astigmatism are not of major concern as long as the focused spot size of the receiver antenna is not larger than the photodetector surface area. For heterodyne and homodyne optical receivers the optical antenna size is limited to the coherence area of the received beam. This subject is discussed further in Chapter 7.

3.3 OTHER OPTICAL COMPONENTS

Other optical components utilized in laser communication systems include phase plates, prisms, polarizers, analyzers, and optical filters.

Phase Plates

A phase plate introduces a specified phase shift between the orthogonal electric field components of an optical wave. Applications include transformation of linear to circular polarization, or vice versa, and compensation for unwanted phase shifts in optical components.

Phase plates are most often constructed of uniaxial crystals. For a wave propagating along a certain direction, perpendicular to the optic axis, these crystals exhibit an "ordinary" index of refraction n_o for wave components of one polarization and an "extraordinary" index of refraction n_e for wave components of the orthogonal polarization (Figure 3-13). The result is that

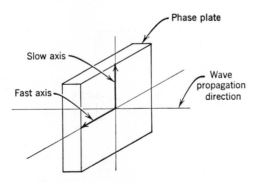

Figure 3-13   Phase plate geometry

the wave components travel at different speeds, and a phase shift is introduced between the wave components. The crystal axis orthogonal to the optic axis which produces the minimum propagation delay of a wave traveling through the crystal is the "fast" axis; the other axis is the "slow" axis. If an optical wave of wavelength λ_c is linearly polarized at 45° with respect to the fast axis, the phase retardation through the crystal of length H_c is

$$\Gamma = \frac{2\pi}{\lambda_c}(n_e - n_o)H_c \tag{3-5}$$

If $\Gamma = \pi$, the crystal is called a half-wave plate, and if $\Gamma = \pi/2$, a quarter-wave plate. In many uniaxial crystals, such as calcite, for a crystal thickness of 1 mm, the phase retardation is actually an integral number of wavelengths plus Γ.

Prisms

Ordinary glass prisms are often used in laser communication systems to deflect light beams. For example, in an optical heterodyne receiver a beam-splitting prism is used to combine orthogonally propagating signal and local oscillator beams into a single parallel beam.

The basic element of a polarization modulation optical receiver is a device called a Wollaston prism. The Wollaston prism, shown in Figure 3-14, consists of two prisms made of biaxial crystals which are cemented together with their optic axes oriented as shown. Incident light which is vertically polarized is deflected in one direction while horizontally polarized light is deflected by the same angle in the opposite direction. Beam-divergence angles of from a few degrees to about 30° are possible. A Wollaston prism may also be employed to combine the carrier and local oscillator beams in an optical heterodyne system [3-20]. The advantage of the Wollaston prism is that none of the carrier energy is lost as is the case when a beam-splitting prism or mirror is used.

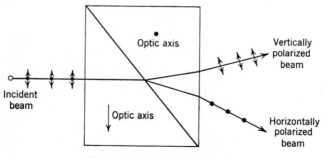

Figure 3-14   Wollaston prism

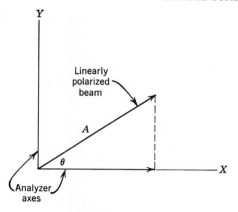

Figure 3-15   *Analyzer operation*

Polarizers and Analyzers

Polarizers and analyzers perform the same physical function and are differentiated by their application. A polarizer transforms an unpolarized light beam into linearly polarized light along the polarizer axis; an analyzer passes only the component of the electric field vector along its axis. If a linearly polarized wave with electric field amplitude A is incident upon a polarizer/analyzer (see Figure 3-15), and the plane of polarization is set at an angle θ with respect to the polarizer/analyzer axis, the amplitude of the exciting beam will be $A \cos \theta$.

The most common type of polarizer/analyzer is the polaroid plate that consists of herapathite crystals embedded in plastic. The crystals have the property of absorbing the electric field component in one direction and passing the orthogonal components [3-6].

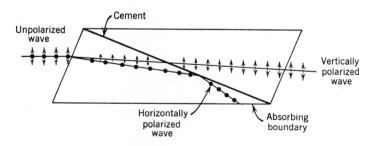

Figure 3-16   *Nicol prism*

Another useful type of polarizer/analyzer is the Nicol prism. It is made of two calcite or quartz prisms held together with transparent cement that has an index of refraction midway between the ordinary and extraordinary ray indices of refraction (Figures 3-16). An unpolarized ray is split into a horizontally polarized ordinary ray and a vertically polarized extraordinary ray. The ordinary ray, deflected by the greater angle, is totally internally reflected at the cement boundary to the side of the crystal where it is absorbed. The extraordinary ray strikes the cement at an angle greater than the critical angle for total internal reflection and passes through the crystal.

Optical Filters

Optical filters are employed before a photodetector to reduce the amount of a background radiation entering the receiver. The most useful types of optical filters are the absorption, scattering, polarization, and interference filters [3-21, 3-22].

Absorption filters are composed of materials that absorb incident energy over certain regions of the optical spectrum. Figure 3-17 shows the transmissivity of several such materials [3-23]. Absorption filters are useful only for gross filtering operations because of their wide bandwidths.

Scattering filters are constructed of optically transparent materials which are finely ground and suspended in a transparent medium of the same refractive index for a given filter center wavelength. At the center wavelength the filter is homogeneous; at all other wavelengths the particles scatter radiation. The bandwidth and center frequency of the filter are dependent upon the refractive indices of the materials, the particle size, and the thickness of the filter. Bandwidths of about 1 micron in the infrared frequencies have been obtained with scattering filters.

If an optical carrier incident upon an optical receiver has a known polarization, a polarization filter may be employed. A polarizer, aligned to the plane of polarization of the optical carrier, is placed before a birefringent crystal phase-retardation plate. The crystal introduces a phase difference between the ordinary and extraordinary ray wave components which is proportional to the optical frequency. The crystal length is set so that the phase difference is a multiple of 2π radians at the center frequency. At all other frequencies, destructive interference takes place. A Lyot filter (Figure 3-18) is a cascaded arrangement of several polarizers interleaved with phase-retardation crystals. Each crystal plate is made twice as thick as the previous one so that it will have twice as many transmission maxima and minima in a given wavelength interval. Figure 3-18 illustrates the transmissivity of each of the phase plate-polarizer combinations and the total filter transmissivity. Lyot filters with a bandpass of 50 Å have been built.

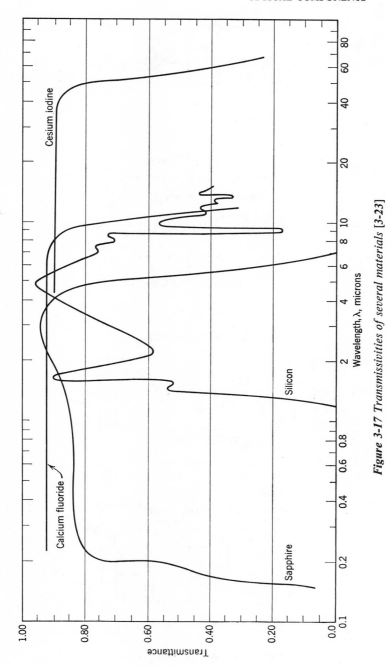

Figure 3-17 Transmissivities of several materials [3-23]

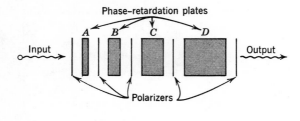

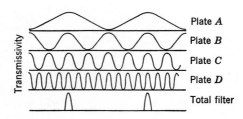

Figure 3-18   Lyot filter

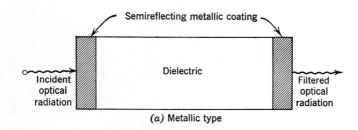

(a) Metallic type

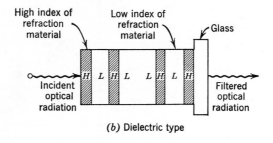

(b) Dielectric type

Figure 3-19   Interference filters

Interference filters provide the narrowest passbands and also can be built for a wide range of center frequencies [3-22]. One type of interference filter is constructed of semireflecting metallic layers deposited on a transparent dielectric, as shown in Figure 3-19a. The metallic layers form a Fabry-Perot interferometer having a narrow passband determined by the reflectivities of the metallic layers, the dielectric refractive index, and the thickness of the filter. Another type of interference filter is the multilayer dielectric filter illustrated in 3-19b. The filter is composed of alternate layers of high and low index of refraction materials deposited on a glass substrate. The optical thickness of the center layer is a half wavelength and the outer layers are each a quarter wavelength thick. The outer *HLH* dielectric stacks act as high reflectance plates of a Fabry-Perot cavity separated by the *LL* dielectric slab. The advantage of the multilayer dielectric filter over the metallic interference filter is that the dielectric plates have higher reflectance and lower absorption than metal plates. Figure 3-20 illustrates the transmission curves of several commercially available interference filters.

REFERENCES

3-1. Lengyl, B. A. *Introduction to Laser Physics.* John Wiley, New York, 1966.

3-2. Yariv, A. *Quantum Electronics.* John Wiley, New York, 1967.

3-3. Smith, W. V and Sorokin, P. P. *The Laser.* McGraw-Hill, New York, 1966.

3-4. Ross, M. *Laser Receivers.* John Wiley, New York, 1966.

3-5. Born, M. and Wolf, E. *Principles of Optics.* Macmillan, New York, 1964.

3-6. Jenkins, F. A. and White, A. E. *Fundamentals of Optics.* McGraw-Hill, New York, 1957.

3-7. Morgan, J. *Introduction to Geometrical and Physical Optics.* McGraw-Hill, New York, 1953.

3-8. Fowles, G. R. *Introduction to Modern Optics.* Holt, Rinehart, and Winston, New, York, 1968.

3-9. Osterink, L. M. and Targ, R. "Single Frequency Light Using the Super-Mode Technique with an Argon FM Laser." *Proceedings of the Symposium on Modern Optics,* J. Fox, ed., XVII, 73–90, 1967.

3-10. Kogelnik, H. and Li, T. "Laser Beams and Resonators." *Proceedings IEEE,* **54** (10), 1312–1329, Oct. 1966.

3-11. Yariv. A. and Gordon, J. P. "The Laser." *Proceedings IEEE,* **51** (1), 4–29, Jan. 1963.

3-12. Bloom, A. L. "Gas Lasers." *Proceedings IEEE,* **54** (10), 1262–1275, Oct. 1966.

3-13. Nathan, M. I. "Semiconductor Lasers." *Proceedings IEEE,* **54** (10), 1276–1290, Oct. 1966.

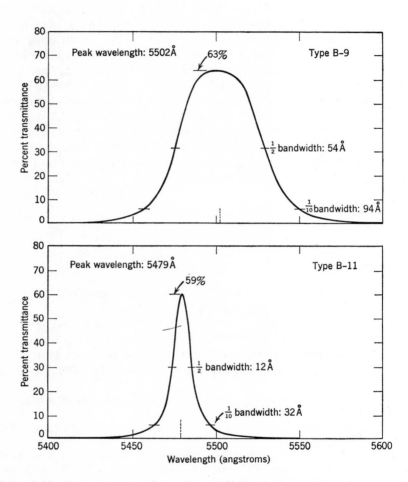

Figure 3-20   *Transmissivity of interference filters (courtesy of Baird-Atomic, Inc.)*

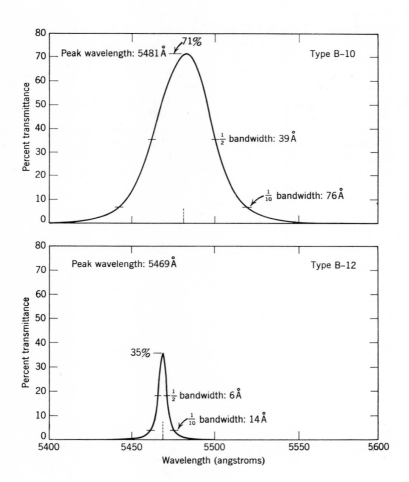

Figure 3-20 (*Continued*)

3-14. Kiss, Z. J. and Pressley, R. J. "Crystalline Solid Lasers." *Proceedings IEEE*, **54** (10) 1236–1248, Oct. 1966.

3-15. Snitzer, E. "Glass Lasers." *Proceedings IEEE*, **54** (10), 1249–1261, Oct. 1966.

3-16. McClung, F. J. and Hellworth, R. W. "Characteristics of Giant Optical Pulsations from Ruby." *Proceedings IEEE*, **51** (1), 46–53, Jan. 1963.

3-17. Ross, M. "YAG Laser Operation by Semiconductor Laser Pumping." *Proceedings IEEE Correspondence*, **56** (2), 196–197, Feb. 1968.

3-18. Bell, L. *The Telescope*. McGraw-Hill, New York, 1922.

3-19. Strong, J. *Concepts of Classical Optics*. W. H. Freeman, London, 1958.

3-20. Siegman, A. E. "Lossless Beam Combination for Optical Heterodyning." *Proceedings IEEE Correspondence*, **52** (1), 94, Jan. 1964.

3-21. Smith, R. A. Jones, F. E., and Chasmar, R. P. *The Detection and Measurement of Infra-red Radiation*, Oxford Press, England, 1957.

3-22. Wolfe, W. L. ed. *Handbook of Military Infrared Technology*. Office of Naval Research, U.S. Government Printing Office, Washington, 1965.

3-23. Rollin, R. A., Jr. and Zwas, F. *Investigation of Space Communications Systems Using Lasers*. Univ. of Michigan, Institute of Science and Technology, RADC–TDR–64–289, Jan. 1966.

chapter 4

MODULATORS

One of the principal potential advantages of laser communications is the high carrier frequency which permits extremely wide-band transmission. Exploitation of this potentiality, of course, requires the development of optical modulators capable of wide-band modulation.

This chapter examines the operating principles of optical modulators from a systems viewpoint. Detailed discussions of modulator materials and construction practices have been omitted due to the breadth of these topics and the rapid evolvement of modulator technology.

4.1 MODULATON MECHANISMS

A large number of mechanisms have been developed for laser modulation. Although many of the mechanisms predate the laser, others are results of directed research arising from the necessity for light weight, low power, broad-band modulators for laser communications.

There are five basic modulation mechanisms: pump power, absorption, spectral, mechano-optic, and electro-optic modulation. Some of these mechanisms are inherently connected with the generation of the laser carrier; others result in discrete modulator assemblies. The discrete modulators may be further categorized as external or internal, depending upon whether they are placed within the laser cavity or outside of it. Internal modulators usually require a lower amount of driving power than external modulators; however, their modulation bandwidths are limited by the bandpass of the laser cavity. Furthermore, internal modulators lower the gain of a laser cavity.

Pump Power Modulation

At some threshold value of laser pumping power, a laser begins to emit a coherent beam, and the beam power increases with increasing pumping power. Thus it is possible to intensity modulate a laser by amplitude modulation of the pump power. For continuous analog modulation, any nonlinearities in the relationship between pumping power and laser power cause distortion.

A gas laser is pump power excited by a direct current or a radio frequency electron discharge through the gas, either of which can be amplitude modulated [4-1]. Regardless of the type of excitation, the electron collision processes provide an inherent modulation rate limitation of about 100 kHz. In the case of radio frequency excitation, the modulation rate must be less than the excitation frequency.

Semiconductor lasers can be pumped by amplitude modulated current sources [4-2]. The modulation rate limitations are the same as the frequency limitations of ordinary diodes. However, the actual practical limitation is often due to the external circuitry problem of switching high currents through the diode at high rates.

It is possible to operate a ruby laser in a semicontinuous mode by providing a flashlamp which operates over a one-second period with a short duty cycle [4-3]. The excitation of the flashlamp is set just below the laser threshold. A second lamp is then periodically activated during the one-second period to produce controlled laser pulses. This "hair-trigger" type of operation is limited to a rate of about 1 kHz.

Absorption Modulation

The optical absorption of certain cystals, such as cadmium selenide, exhibits a sharp, low frequency cutoff which can be controlled by application

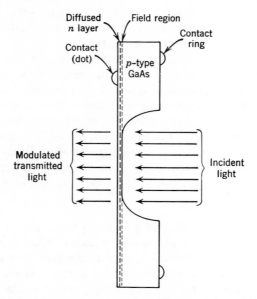

Figure 4-1   p-n junction edge effect modulator [4-7]

of an electric field [4-4]. This result is called the Franz-Keldysh effect [4-5]. Shifting the absorption edge slightly results in large changes in the transmissivity of the crystal.

The optical absorption at a *p-n* semiconductor junction can also be shifted by an electric field [4-6, 4-7]. Figure 4-1 illustrates a *p-n* junction edge effect modulator. For both crystals and semiconductors, modulation may be performed up to microwave rates, but the absorption modulator is limited to low power transmitters because of power dissipation in the absorbing materials.

Spectral Modulation

The oscillation frequency of a laser is adjustable over a narrow range by changes in the length of the cavity. One convenient method of cavity length modulation is to mount the cavity end mirrors on a magnetostrictive material which changes its length proportional to an applied electric field [4-8]. The resonant frequency, f_R, of the cavity is

$$f_R = \frac{nc}{2L_c} \tag{4-1}$$

where n is an integer, c is the velocity of light, and L_c is the cavity length.

A change in cavity length, ΔL, produces a change in resonant frequency of

$$\Delta f_R = \frac{nc}{2L_c} - \frac{nc}{2(L + \Delta L_c)} \tag{4-2}$$

For small changes in length

$$\Delta f_R \approx \frac{nc \, \Delta L_c}{2L_c{}^2} = f_R\left(\frac{\Delta L_c}{L_c}\right) \tag{4-3}$$

The modulation bandwidth is limited to the spacing between modes (e.g., 150 MHz with 1-meter cavity) for a multimode laser. Extraneous mechanical movements of the mirror can easily lead to random frequency shifts of 100 kHz, providing a lower limit of frequency modulation. A 1 MHz frequency shift at a wavelength of 0.5 micron (6×10^{14} Hz) requires a cavity length change of $\Delta L_c/L_c = 1.67 \times 10^{-9}$ which is achievable with magnetostrictive materials. The maximum modulation rate at which the line frequency can be shifted is limited to about 100 MHz by the time response of the magnetostrictive material.

A magnetic field applied to certain solid or gaseous materials, through which a laser beam is propagating, results in the splitting of the normal laser carrier frequency into components separated from the carrier frequency by plus or minus a difference frequency which is proportional to the field strength

[4-9]. This is called the Zeeman effect and results in a frequency shift of [4-10]

$$\Delta f_z = \frac{qH}{4\pi mc} \tag{4-4}$$

where m is the mass of an electron and H is the magnetic field strength (Oersted).

A splitting of the laser emission frequency, called the Stark effect, also occurs when the laser beam passes through gaseous or solid materials and is subjected to an electric field. The frequency shift is [4-10]

$$\Delta f_s = 0.5 \times 10^6 E \tag{4-5}$$

where E is the electric field strength (volts/cm).

Large magnetic and electric fields are required to obtain significant frequency shifts. The modulation bandwidth is limited by the rate at which the fields can be switched. Zeeman or Stark frequency shifting provides a convenient method of slowly tuning a local oscillator laser for heterodyning.

Mechano-Optic Modulation

Certain crystals and liquids exhibit changes in index of refraction due to mechanical strain. If the strains are mechanically applied, the refractive index change is called the photoelastic effect. Index changes are also caused by the piezoelectric effect by which an electric field applied to a material causes it to deform mechanically. Ultrasonic waves propagating through a material produce mechanical strains which change the refractive index due to the acousto-optic effect [4-11 to 4-13]. Mechanically induced changes in the refractive index of a medium produce centers of diffraction for incident light. The intensity and frequency of the diffracted light are proportional to the strain which, in turn, is proportional to a modulation signal.

Photoelastic and piezoelectric modulators have restricted modulation bandwidths due to the mechanical frequency response of the modulator materials. The frequency limitation of an acousto-optic modulator is the transit time of the ultrasonic wave across the carrier beam. By focusing the laser beam to a narrow diameter, reasonably large modulation bandwidths are possible. The acousto-optic modulator is of considerable interest because it offers a means of laser frequency shifting which is useful for tuning a local oscillator.

Figure 4-2 illustrates the basic configuration of an acousto-optic modulator [4-14]. With an unmodulated ultrasonic wave propagating through the medium (which may be quartz, cadmium sulphide, lithium niobate, or simply water), sinusoidal variations in the refractive index, n, are produced with a wavelength equal to the ultrasonic wavelength, λ_s. A narrow collimated laser

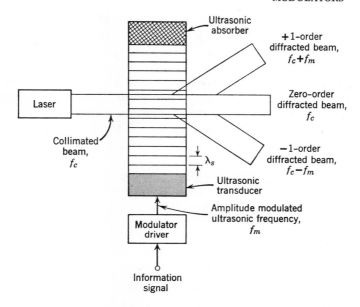

Figure 4-2 *Basic acousto-optic modulator*

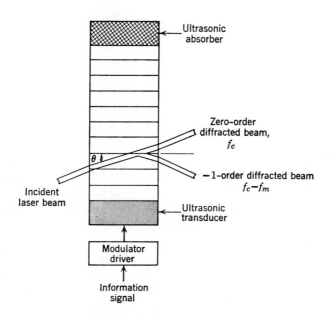

Figure 4-3 *Bragg angle acousto-optic modulator*

beam striking the medium is scattered into the primary diffraction orders shown in Figure 4-2. Higher diffraction orders are also produced, but their intensities can be made negligible. The frequency of the zero-order beam is the same as the incident beam f_c, while the frequencies of the plus one- and minus one-orders are $f_c + f_m$ and $f_c - f_m$. Thus, frequency modulation of the ultrasonic wave results in frequency modulation of the diffracted beams [4-15]. Amplitude modulation of the ultrasonic wave causes a corresponding intensity modulation on each of the diffracted beams. Only one of the three diffracted beams is useful for purposes of communication; the others may be eliminated by a field stop.

The inherent inefficiency resulting from the stopping of two of the diffraction beams is overcome by the use of a Bragg angle acousto-optic modulator [4-14]. If the laser beam strikes the ultasonic cell, as shown in Figure 4-3, at the Bragg angle defined by

$$\theta \equiv \sin^{-1}\left(\frac{\lambda_c}{2\lambda_s}\right) \tag{4-6}$$

where λ_c and λ_s are the laser and ultrasonic wavelengths, the plus one diffracted wave intensity is reduced to zero. The intensities of the zero- and minus one-order beams are

$$\mathscr{I}_0 = \cos^2\left(\frac{\nu}{2}\right) \tag{4-7}$$

$$\mathscr{I}_{-1} = \sin^2\left(\frac{\nu}{2}\right) \tag{4-8}$$

where ν, the Raman-Nath parameter [4-16], is

$$\nu \equiv \frac{2\pi w_u \, \Delta n}{\lambda_c} \tag{4-9}$$

and w_u is the ultrasonic cell width and Δn is the change in the index of refraction of the medium caused by the ultrasonic wave. The value of Δn is a function of the normal index of refraction, the electro-optic tensor, and the strain tensor; the exact relationship is dependent upon the characteristics of the medium and the geometrical relations between the directions of propagation of the sound and light waves.

For sufficiently large values of the Raman-Nath parameter, under ideal experimental conditions, approximately 95% of the intensity of the incident laser beam may be transferred to the minus one diffraction order. The Raman-Nath parameter is inversely proportional to wavelength; hence, acousto-optic modulators are not as effective at higher wavelengths. Frequency modulation is possible with the Bragg angle modulator by transferring the carrier wave energy to the minus one diffraction order and then frequency modulating the ultrasonic wave [4-17].

Electro-Optic and Magneto-Optic Modulation

Certain crystals and liquids become birefringent in the presence of an electric or magnetic field [4-18, 4-19]. As a result, the index of refraction of the medium is electrically or magnetically controllable along the coordinate axes of the medium.

If a magnetic field is applied to a magneto-optic liquid such as nitro-benzene, the molecules, which are magnetically and optically anisotropic, will line up in the direction of the applied field. This phenomenon is called the Cotton-Mouton magneto-optic effect. The electric field analog is called the Kerr electro-optic effect. Neither effect is widely used for optical modulators because of the difficulty in handling the liquids which are generally caustic and poisonous. Also, application of fields to these liquids for a prolonged period often causes decomposition.

The electro-optic effect in crystals, called the Pockel's electro-optic effect, is the basis for many electro-optic modulators. Crystals such as potassium dihydrogen phosphate (KH_2PO_4 or, more commonly, KDP), ammonium dihydrogen phosphate ($NH_4H_2PO_4$ or ADP), cuprous chloride (CuCl), and lithium niobate ($LiNbO_3$), which are transparent at the visible wavelengths, and gallium arsenide (GaAs), which is transparent to infrared frequencies, exhibit the Pockel's effect. References [4-20] to [4-22] list the characteristics of the electro-optic crystals.

In an electro-optic crystal the index of refraction, n, along a crystal axis may be written in series form as

$$\frac{1}{n^2} = \frac{1}{n_w^2} + rE + RE^2 + \cdots \qquad (4\text{-}10)$$

where E is the applied electric field, n_w is the index of refraction without the electric field, r is the linear electro-optic coefficient, and R is the quadratic electro-optic coefficient. In crystals that are useful for electro-optic modulation, either the linear or quadratic coefficient is usually predominant; and the change in the applied electric field produces a linear or quadratic change in the index of refraction.

The indices of refraction along the coordinate axes (X, Y, Z) of an electro-optic crystal without an electric field are related by the following equation called the optical indicatrix or index ellipsoid.

$$\left(\frac{X}{n_X}\right)^2 + \left(\frac{Y}{n_Y}\right)^2 + \left(\frac{Z}{n_Z}\right)^2 = 1 \qquad (4\text{-}11)$$

Figure 4-4a illustrates the optical indicatrix of an electro-optic crystal with no applied field. In the presence of an electric field the equation for the ellipsoid, related to the crystallographic axes, is given by

$$\left(\frac{X}{n_X}\right)^2 + \left(\frac{Y}{n_Y}\right)^2 + \left(\frac{Z}{n_Z}\right)^2 + \sum_{i,j,k,p=1}^{3} [r_{ijk}E_{\zeta_k} + R_{ijkp}E_{\zeta_k}E_{\zeta_p}]\zeta_i\zeta_j = 1 \quad (4\text{-}12)$$

where $\zeta_1 \equiv X$, $\zeta_2 \equiv Y$, and $\zeta_3 \equiv Z$. In general the effects of the electric field are to rotate the ellipsoid with respect to the crystallographic axis and to change its shape. Figure 4-4b shows the optical indicatrix of an electro-optic crystal subject to an electric field. In this example, the ellipsoid is simply rotated about the Z axis of the crystal.

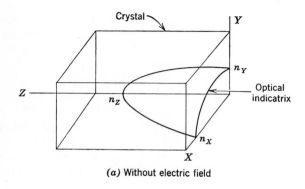

(a) Without electric field

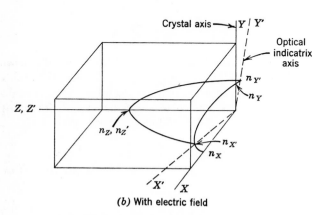

(b) With electric field

Figure 4-4   *Optical indicatrix geometry*

A notational shorthand has been developed whereby the indices i, j and k, p in Equation 4-12 can be written as

i	j	m
k	p	n
1	1	1
2	2	2
3	3	3
2	3	4
3	2	4
1	3	5
3	1	5
1	2	6
2	1	6

With this notation there are 18 linear and 36 quadratic electro-optic coefficients. Symmetry of the crystal causes many of the r_{mk} and R_{mn} electro-optic coefficients to be zero or identical. Mason has compiled a listing of electro-optic coefficients for the various crystal classes [4-23].

When considering only the linear electro-optic effect, the equation for the index ellipsoid can be written as

$$\left(\frac{X}{n_X}\right)^2 + \left(\frac{Y}{n_Y}\right)^2 + \left(\frac{Z}{n_Z}\right)^2 + \sum_{k=1}^{3} [r_{1k}X^2 + r_{2k}Y^2 + r_{3k}Z^2 + 2r_{4k}YZ$$

$$+ 2r_{5k}XZ + 2r_{6k}XY]E_{\zeta_k} = 1 \quad (4\text{-}13)$$

There exists another coordinate system (X', Y', Z') containing the principal axes of the rotated ellipsoid for which the equation of the optical indicatrix is

$$\left(\frac{X'}{n_{X'}}\right)^2 + \left(\frac{Y'}{n_{Y'}}\right)^2 + \left(\frac{Z'}{n_{Z'}}\right)^2 = 1 \quad (4\text{-}14)$$

The indices of refraction $n_{X'}$, $n_{Y'}$, and $n_{Z'}$ are related to the indices n_X, n_Y, and n_Z through a coordinate transformation described by the direction cosines α_i, β_i, and γ_i given below

	X'	Y'	Z'
X	α_1	α_2	α_3
Y	β_1	β_2	β_3
Z	γ_1	γ_2	γ_3

By this transformation and considering only the linear electro-optic effect [4-24],

$$\left(\frac{1}{n_{X'}}\right)^2 = \left(\frac{1}{n_X}\right)^2 \alpha_1{}^2 + \left(\frac{1}{n_Y}\right)^2 \alpha_2{}^2 + \left(\frac{1}{n_Z}\right)^2 \alpha_3{}^2$$

$$+ \sum_{k=1}^{3} [r_{1k}\alpha_1{}^2 + r_{2k}\alpha_2{}^2 + r_{3k}\alpha_3{}^2 + 2r_{4k}\alpha_2\alpha_3$$

$$+ 2r_{5k}\alpha_1\alpha_3 + 2r_{6k}\alpha_1\alpha_2]E_{\zeta_k} \quad (4\text{-}15)$$

$$\left(\frac{1}{n_{Y'}}\right)^2 = \left(\frac{1}{n_X}\right)^2 \beta_1{}^2 + \left(\frac{1}{n_Y}\right)^2 \beta_2{}^2 + \left(\frac{1}{n_Z}\right)^2 \beta_3{}^2$$

$$+ \sum_{k=1}^{3} [r_{1k}\beta_1{}^2 + r_{2k}\beta_2{}^2 + r_{3k}\beta_3{}^2 + 2r_{4k}\beta_2\beta_3$$

$$+ 2r_{5k}\beta_1\beta_3 + 2r_{6k}\beta_1\beta_2]E_{\zeta_k} \quad (4\text{-}16)$$

$$\left(\frac{1}{n_{Z'}}\right)^2 = \left(\frac{1}{n_X}\right)^2 \gamma_1{}^2 + \left(\frac{1}{n_Y}\right)^2 \gamma_2{}^2 + \left(\frac{1}{n_Z}\right)^2 \gamma_3{}^2$$

$$+ \sum_{k=1}^{3} [r_{1k}\gamma_1{}^2 + r_{2k}\gamma_2{}^2 + r_{3k}\gamma_3{}^2 + 2r_{4k}\gamma_2\gamma_3$$

$$+ 2r_{5k}\gamma_1\gamma_3 + 2r_{6k}\gamma_1\gamma_2]E_{\zeta_k} \quad (4\text{-}17)$$

If an optical wave travels along the Z' axis of a crystal and the wave is linearly polarized at 45° to the X' axis, the wave will experience a phase retardation, Γ, between its X' and Y' components equal to

$$\Gamma = \frac{2\pi H_c}{\lambda_c}(n_{X'} - n_{Y'}) \quad (4\text{-}18)$$

where H_c is the crystal length. Techniques for determining the index ellipsoid and phase retardation of an electro-optic crystal are illustrated by the following examples.

Tetragonal crystals of crystal class $\overline{4}2m$: Members of this crystal class are KDP and ADP [4-24]. Consider such a crystal with an optical wave traveling along the Z axis and an electric field in the same direction, as shown in Figure 4-5. The coincident field and beam directions can be realized by the use of holes in the electrodes or optically transparent electrodes.

In these crystals the quadratic electro-optic coefficient is negligible; due to crystal symmetries, the only nonzero linear electro-optic coefficients are $r_{41} = r_{52}$ and r_{63}. The optical indicatrix then reduces to

$$\left(\frac{X}{n_o}\right)^2 + \left(\frac{Y}{n_o}\right)^2 + \left(\frac{Z}{n_e}\right)^2 + 2r_{63}E_Z XY = 1 \quad (4\text{-}19)$$

where n_o and n_e are called the ordinary ray and extraordinary ray indices of

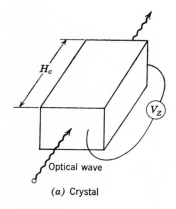

(a) Crystal

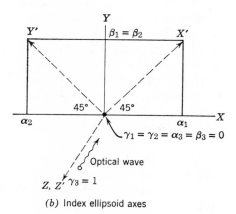

(b) Index ellipsoid axes

Figure 4-5 *Crystal geometry of class* $\bar{4}2m$ *crystal with Z axis electric field*

refraction, respectively. This equation indicates that the Z and Z' axes coincide and that the index ellipsoid is rotated by an angle of 45° with respect to the crystallographic axis. Hence, from Figure 4-5, the direction cosines are $\gamma_3 = 1$, $\gamma_1 = \gamma_2 = \alpha_3 = \beta_3 = 0$, $\alpha_1 = \beta_2 = \cos(\pi/4)$, and $-\alpha_2 = \beta_1 = \sin(45°)$. Then, from Equations 4-15 and 4-16 the indices of refraction along the X' and Y' coordinates are

$$\frac{1}{n_{X'}^2} = \frac{1}{n_o^2} - r_{63}E_Z \tag{4-20}$$

$$\frac{1}{n_{Y'}^2} = \frac{1}{n_o^2} + r_{63}E_Z \tag{4-21}$$

To a good approximation

$$n_{X'} \approx n_o + \tfrac{1}{2}n_o{}^3 r_{63} E_Z \qquad (4\text{-}22)$$

$$n_{Y'} \approx n_o - \tfrac{1}{2}n_o{}^3 r_{63} E_Z \qquad (4\text{-}23)$$

The phase retardation for a wave traveling along the $Z = Z'$ axis and linearly polarized along the X axis (i.e., $+45°$ to the Y' axis) is then

$$\Gamma = \frac{2\pi n_o{}^3 r_{63} V_Z}{H_c} \qquad (4\text{-}24)$$

where $V_Z = E_Z H_c$ is the voltage across the crystal. Hence, for crystals of the $\overline{4}2m$ class with a Z-axis field and with proper orientation, the phase retardation is independent of the crystal length. For a KDP crystal $n_o = 1.51$, $r_{63} = 1.03 \times 10^{-11}$ meters/volt, and the voltage required to make $\Gamma = \pi$ radians (the half-wave voltage) is 11 kv at a wavelength of 0.63 micron. Crystals may be cascaded to reduce the driving voltage.

Cubic crystals of crystal class $\overline{4}3m$: Member of this class are cuprous chloride (CuCl) and zinc sulphide (ZnS) [4-25]. Crystal symmetries reduce the nonzero electro-optic coefficients to $r_{41} = r_{52} = r_{63}$. If an electric field is applied along the Y axis transverse to the direction of beam propagation, as shown in Figure 4-6, the optical indicatrix is [4-26]

$$\left(\frac{X}{n_o}\right)^2 + \left(\frac{Y}{n_o}\right)^2 + \left(\frac{Z}{n_o}\right)^2 + 2r_{41}E_X X_Z = 1 \qquad (4\text{-}25)$$

The ellipsoid is rotated about the $Y = Y'$ axis by an angle of $45°$. From Figure 4-6 the direction cosines are $\beta_2 = 1$, $\alpha_2 = \beta_1 = \beta_3 = \gamma_2 = 0$, $\alpha_1 = \gamma_3 = \cos(\pi/4)$, and $\alpha_3 = -\gamma_1 = \sin(\pi/4)$. The indices of refraction along the X' and Y' axes then reduce to

$$\left(\frac{1}{n_{X'}}\right)^2 = \left(\frac{1}{n_o}\right)^2 + r_{41}E_Y$$

$$\left(\frac{1}{n_{Y'}}\right)^2 = \left(\frac{1}{n_o}\right)^2$$

If an optical wave linearly polarized at $+45°$ to the X' axis travels down the Z' axis the phase retardation will be

$$\Gamma = \frac{\pi n_o{}^3 r_{41} V_Y}{\lambda_c}\left(\frac{H_c}{D_c}\right) \qquad (4\text{-}26)$$

where D_c is the thickness of the crystal in the Y dimension. Hence, the phase retardation can be increased for a given applied voltage by increasing the length to thickness ratio of the crystal.

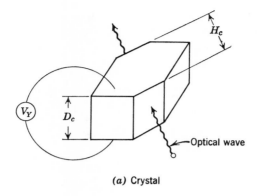

(a) Crystal

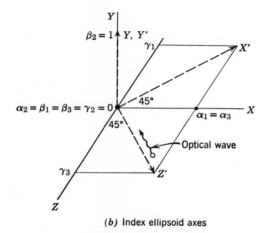

(b) Index ellipsoid axes

Figure 4-6   *Crystal geometry of class* $\bar{4}3m$ *crystal with X axis field*

Hexagonal crystals of crystal class 6: In this class of crystals the nonzero electro-optic coefficients are

$$r_{12} = r_{61} = -r_{22}$$

$$r_{21} = r_{62} = -r_{11}$$

and the index ellipsoid in the $Z = 0$ plane for fields along the X and Y axes is [4-27]

$$\left[\left(\frac{1}{n_o}\right)^2 + (r_{11}E_X - r_{22}E_Y)\right]X^2 + \left[\left(\frac{1}{n_o}\right)^2 + (-r_{11}E_X + r_{22}E_Y)\right]Y^2$$

$$+ 2[-r_{22}E_X - r_{11}E_Y]XY = 1 \quad (4\text{-}27)$$

If the electric fields are $E_X = E \cos \xi$ and $E_Y = E \sin \xi$, the index ellipsoid will be rotated by an angle

$$\theta = \frac{1}{2} \left\{ \xi + \sin^{-1} \left[\frac{r_{22}}{\sqrt{r_{11}^2 + r_{22}^2}} \right] \right\}$$

about the $Z = Z'$ axis. Along its principal axes the equation for the ellipsoid in the $Z = 0$ plane is

$$\left[\left(\frac{1}{n_o} \right)^2 + \sqrt{r_{11}^2 + r_{22}^2}\, E \right] X'^2 + \left[\left(\frac{1}{n_o} \right)^2 - \sqrt{r_{11}^2 + r_{22}^2}\, E \right] Y'^2 = 1 \quad (4\text{-}28)$$

The phase retardation for an optical wave traveling along the Z' axis is approximately

$$\Gamma = \frac{2\pi L n_o^3 \sqrt{r_{11}^2 + r_{22}^2}\, E}{\lambda_c} \quad (4\text{-}29)$$

Summary of Modulation Mechanisms

Table 4-1 lists the modulation methods that may be obtained with various modulation mechanisms. The most practical and widely used modulation mechanisms have been found to be pump power modulation of a semiconductor laser, acousto-optic modulation, and electro-optic modulation.

Modulation Mechanism	Modulation Method				
	AM	IM	FM	PM	PL
Gas laser pump power		×			
Semiconductor laser pump power		×			
Solid-state laser pump power		×			
Absorption edge		×			
Cavity length change			×		
Zeeman			×		
Stark			×		
Photoelastic		×	×		
Piezoelectric		×	×		
Acousto-optic		×	×		
Magneto-optic	×		×	×	×
Electro-optic	×	×	×	×	×

Table 4-1. Modulation Methods Possible with Various Modulation Mechanisms

The next section considers electro-optic modulators in greater detail. Problems associated with the construction of wide-bandwidth modulators are discussed in references [4-28] to [4-34].

4.2 ELECTRO-OPTIC MODULATORS

Electro-optic modulators have gained wide acceptance for laser communication and allied applications because: (a) all forms of modulation are possible, (b) wide-band operation can be achieved, (c) modulation is possible at all optical wavelengths, and (d) electro-optic modulators are not overly expensive or difficult to construct. In the following discussion examples are given of intensity, polarization, and frequency electro-optic modulators. The examples are not restricted to any particular crystal materials but merely specify operation of the modulator in terms of the relation between the applied voltage and phase retardation.

Electro-Optic Intensity Modulator

A diagram of an electro-optic intensity modulator is shown in Figure 4-7. The Pockel's cell produces a phase retardation, Γ, linearly proportional to an applied voltage.

A laser beam of intensity $\mathcal{I}_s$ is linearly polarized at an angle of 45° with respect to the crystal index ellipsoid, and an analyzer following the modulator is set orthogonal to the laser polarization. Thus, with no field on the crystal, the laser beam is completely attenuated by the modulator. The laser

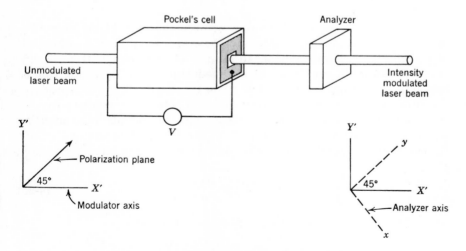

Figure 4-7 *Electro-optic intensity modulator*

polarization state, Pockel's cell, and the analyzer may be represented by polarization matrices referenced to the crystallographic axis [Appendix A].

The polarization matrix of the laser beam is

$$\mathscr{L} = \sqrt{\frac{\mathscr{I}_s}{2}} \begin{bmatrix} 1 \\ 1 \end{bmatrix} \tag{4-30}$$

The modulator with a total phase retardation of Γ has an operational matrix

$$M_M = \begin{bmatrix} e^{i\Gamma/2} & 0 \\ 0 & e^{-i\Gamma/2} \end{bmatrix} \tag{4-31}$$

and the analyzer matrix operator is

$$M_A = \frac{1}{2} \begin{bmatrix} 1 & -1 \\ -1 & 1 \end{bmatrix} \tag{4-32}$$

The modulator output beam polarization matrix is then

$$\mathscr{L}_o = M_A M_M \mathscr{L} = \sqrt{\frac{\mathscr{I}_s}{2}} i \sin\frac{\Gamma}{2} \begin{bmatrix} 1 \\ -1 \end{bmatrix} \tag{4-33}$$

which represents light linearly polarized at $-45°$ with respect to the modulator crystal X' axis and whose electric field amplitude is proportional to $\sin(\Gamma/2)$. The imaginary term, i, represents a $90°$ phase shift of both the X' and Y' electric field components entering the modulator and is unimportant in this application. The modulator output intensity is

$$\mathscr{I}_{so} = \mathscr{I}_s \sin^2\left(\frac{\Gamma}{2}\right) \tag{4-34}$$

Figure 4-8 illustrates the relationship between the ratio of the output-to-input intensity of the modulator and the phase retardation Γ. Since the phase retardation is proportional to the modulating voltage, there is considerable distortion for linear analog modulation. The distortion can be minimized by biasing the crystal with a constant electric field which produces a quarter wave of phase retardation ($\Gamma = \pi/2$) with no modulating voltage. A more desirable technique is to optically bias the modulator by inserting a quarter-wave phase-retardation plate in the light path before or after the Pockel's cell. The operational matrix of the quarter-wave phase-retardation plate is

$$M_Q = \begin{bmatrix} e^{i\pi/4} & 0 \\ 0 & e^{-i\pi/4} \end{bmatrix} \tag{4-35}$$

and the resulting modulator output polarization matrix is

$$\mathscr{L}_o = M_A M_M M_Q \mathscr{L} = \sqrt{\frac{\mathscr{I}_s}{2}} i \sin\left(\frac{\Gamma}{2} + \frac{\pi}{4}\right) \begin{bmatrix} 1 \\ -1 \end{bmatrix} \tag{4-36}$$

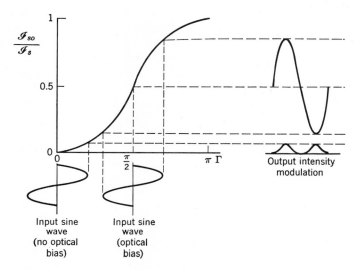

Figure 4-8 *Relationship between sine wave input and intensity output of electro-optic modulator*

The modulator output intensity then becomes

$$\mathcal{I}_{so} = \mathcal{I}_s[\tfrac{1}{2} + \tfrac{1}{2} \sin \Gamma] \qquad (4\text{-}37)$$

Thus, for no modulating voltage the modulator output intensity is half the input intensity.

In the electro-optic crystal the phase retardation is linearly related to the driving voltage. For an applied sine wave modulating voltage, Γ may be written as

$$\Gamma = K_m V_m \sin \omega_m t \qquad (4\text{-}38)$$

where K_m is a constant of proportionality, V_m is the maximum modulation voltage, and $\omega_m/2\pi$ is the modulation frequency. Then the modulator output intensity is

$$\mathcal{I}_{so} = \frac{\mathcal{I}_s}{2} [1 + \sin (K_m V_m \sin \omega_m t)] \qquad (4\text{-}39)$$

Equation 4-39 can be expanded in terms of Bessel functions of the first kind to yield

$$\mathcal{I}_{so} = \frac{\mathcal{I}_s}{2} + \mathcal{I}_s[J_1(K_m V_m) \sin \omega_m t + J_3(K_m V_m) \sin 3\omega_m t$$
$$+ J_5(K_m V_m) \sin 5\omega_m t + \cdots] \qquad (4\text{-}40)$$

The modulator output thus consists of a bias term of one-half the modulator input intensity, the desired fundamental frequency term of relative amplitude $J_1(K_m V_m)$, and odd harmonics of the fundamental. These harmonic terms cause distortion in the modulation process. The amount of distortion, $\mathscr{D}_i$, is defined to be the ratio of the root mean square of the harmonic amplitudes to the fundamental term amplitude.

$$\mathscr{D}_i \equiv \frac{\sqrt{[J_3(K_m V_m)]^2 + [J_5(K_m V_m)]^2 + \cdots}}{J_1(K_m V_m)} \times 100\% \qquad (4\text{-}41)$$

A modulation index, M_{IM}, describing the peak excursion of the modulator output intensity about the mean output intensity, is defined as (Equation 2-26)

$$M_{\mathrm{IM}} = \frac{(\mathscr{I}_{so})_{\max} - \frac{1}{2}\mathscr{I}_s}{\frac{1}{2}\mathscr{I}_s} \times 100\% \qquad (4\text{-}42)$$

If the harmonic distortion terms are neglected, the modulation index becomes

$$M_{\mathrm{IM}} = 2J_1(K_m V_m) \qquad (4\text{-}43)$$

To achieve a modulation index approaching 100%, the third and higher harmonic distortion components become large. Figure 4-9 illustrates the third harmonic distortion versus modulation percentage [4-35].

Electro-Optic Polarization Modulator

In the electro-optic polarization modulator, a laser beam of intensity $\mathscr{I}_s$ is linearly polarized at an angle of 45° with respect to the crystal axes of a Pockel's cell which provides a phase retardation proportional to the applied

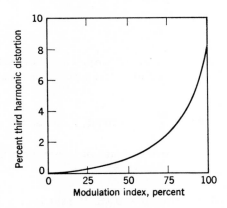

Figure 4-9   *Comparision of modulation index and percentage of third harmonic distortion*

voltage. At the output of the modulator the polarization matrix is

$$\mathscr{L}_o = M_M \mathscr{L} = \sqrt{\frac{\mathscr{I}_s}{2}} \begin{bmatrix} e^{i\Gamma/2} \\ e^{-i\Gamma/2} \end{bmatrix} \tag{4-44}$$

For a positive quarter-wave phase shift of $\Gamma = \pi/2$

$$\mathscr{L}_o = \sqrt{\frac{\mathscr{I}_s}{2}} e^{-i\pi/4} \begin{bmatrix} i \\ 1 \end{bmatrix} \tag{4-45}$$

and left circularly polarized light is obtained. For a negative quarter-wave phase shift of $\Gamma = -\pi/2$

$$\mathscr{L}_o = \sqrt{\frac{\mathscr{I}_s}{2}} e^{i\pi/4} \begin{bmatrix} -i \\ 1 \end{bmatrix} \tag{4-46}$$

and right circularly polarized light results. Phase retardation between $\Gamma = -\pi/2$ and $\Gamma = +\pi/2$ results in elliptical polarization with the degree of ellipticity directly proportional to the phase retardation.

Electro-Optic Frequency Modulator

Figure 4-10 contains a diagram of an electro-optic frequency modulator. The Pockel's cell is a crystal of threefold rotational symmetry, and the orthogonal driving voltages are $V_X = V \cos \xi$ and $V_Y = V \sin \xi$. Under

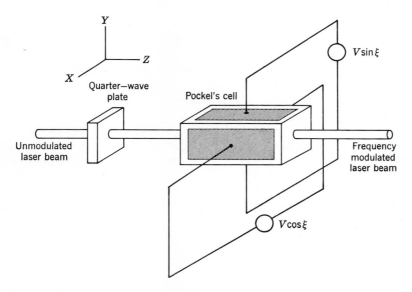

Figure 4-10 *Electro-optic frequency modulator*

these conditions the phase retardation Γ is linearly related to V. The index ellipsoid of the crystal is rotated by an angle θ (with respect to the crystallographic axis) which is linearly proportional to ξ. The modulator operational matrix referred to the crystal axis is found by the transformation (Appendix B)

$$M_M' = M_C M_M M_C^\dagger = \begin{bmatrix} \cos\theta & -\sin\theta \\ \sin\theta & \cos\theta \end{bmatrix} \begin{bmatrix} e^{i\Gamma/2} & 0 \\ 0 & e^{-i\Gamma/2} \end{bmatrix} \begin{bmatrix} \cos\theta & \sin\theta \\ -\sin\theta & \cos\theta \end{bmatrix}$$

(4-47)

or, in expanded form,

$$M_M' = \begin{bmatrix} e^{i\Gamma/2}\cos^2\theta + e^{-i\Gamma/2}\sin^2\theta & e^{i\Gamma/2}\sin\theta\cos\theta - e^{-i\Gamma/2}\sin\theta\cos\theta \\ e^{i\Gamma/2}\sin\theta\cos\theta - e^{-i\Gamma/2}\sin\theta\cos\theta & e^{i\Gamma/2}\sin^2\theta + e^{-i\Gamma/2}\cos^2\theta \end{bmatrix}$$

(4-48)

A laser beam of intensity $\mathscr{I}_s$ is linearly polarized at 45° to the reference axis and passed through a quarter-wave plate. The resultant beam, which is incident on the modulator, is right circularly polarized with polarization matrix

$$\mathscr{L} = \sqrt{\frac{\mathscr{I}_s}{2}} \begin{bmatrix} i \\ 1 \end{bmatrix} e^{i\omega t}$$

(4-49)

where the usually implicit time function $e^{i\omega t}$ has been included. The polarization matrix of the light leaving the modulator, after some trigonometric manipulation, is found to be

$$\mathscr{L}_o = M_M'\mathscr{L} = \sqrt{\frac{\mathscr{I}_s}{2}}\cos\left(\frac{\Gamma}{2}\right)e^{i\omega t}\begin{bmatrix} i \\ 1 \end{bmatrix} - \sqrt{\frac{\mathscr{I}_s}{2}}\,i\sin\left(\frac{\Gamma}{2}\right)e^{i(\omega t - 2\theta)}\begin{bmatrix} -i \\ 1 \end{bmatrix}$$
(4-50)

The modulator output light is thus composed of two beams, one right circularly polarized at the carrier frequency with amplitude proportional to $\cos{(\Gamma/2)}$ and the other left circularly polarized at a frequency $(\omega - 2d\theta/dt)$ and with amplitude proportional to $\sin{(\Gamma/2)}$.

For analog frequency modulation, Γ is set to π radians to place all the laser carrier power in the right circularly polarized component. Then frequency modulation of the driving voltage directly produces frequency modulation of the optical carrier. Frequency shift keying of the optical carrier is performed by setting ξ to give the desired frequency separation between the transmitted frequencies and then setting $\Gamma = 0$ or π to represent "ones" and "zeros."

REFERENCES

4-1. Schiel, E. J. and Bolmarcich, J. J. "Direct Modulation of a He–Ne Gas Laser." *Proceedings IEEE Correspondence*, **51** (6), 940–941, June 1963.

4-2. Davison, L. A. "Pulse Modulation of Gallium Arsenide Injection Luminescent Diode Laser." *Proceedings IEEE Correspondence*, **51** (8), 1141–1142, Aug. 1963.

4-3. Stitch, M. L., Woodbury, E. J., and Morse, J. H. "Repetitive Hair-Trigger Mode of Optical Maser Operation." *Proceedings IRE*, **49** (10), 1571–1572, Oct. 1961.

4-4. Williams, R. "Electrical Field Induced Light Absorption in CdS." *Physical Review*, **117**, 1487–1490, Mar. 1960.

4-5. Keldysh, L. V. "The Effect of a Strong Electric Field on the Optical Properties of Insulating Crystals." *Journal of Experimental and Theoretical Physics*, **34**, 1138–1141, May 1958.

4-6. Renton, C. A. "Amplitude Modulation of Light by Reverse Bias *p-n* Junctions." *Proceedings IEEE Correspondence*, **52** (1), 93–94, Jan. 1964.

4-7. Racette, G. "Absorption Edge Modulator Utilizing a *p-n* Junction." *Proceedings IEEE Correspondence*, **52** (6), 716, June 1964.

4-8. Benet, W. R. et al. "Magnetostricitively Tuned Optical Maser." *Review Scientific Instruments*, **33**, 601–605, June 1962.

4-9. Schmidt, B. M., Williams, J. M., and Williams, D. "Magnetic-Optic Modulation of a Light Beam in Sodium Vapor." *Journal Optical Society of America*, **54** (4), 454–455, April 1964.

4-10. Jenkins, F. A. and White, H. E. *Fundamentals of Optics*, McGraw-Hill, New York, 1957.

4-11. Quate, C. F., Wilkinson, C. D. W., and Winslow, D. K. "Interaction of Light and Microwave Sound." *Proceedings IEEE*, **53** (10), 1604–1622, Oct. 1965.

4-12. Gordon, E. I. "A Review of Acousto-optical Deflection and Modulation Devices." *Proceedings IEEE*, **54** (10), 1391–1401, Oct. 1966.

4-13. Klein, W. R. and Cook, B. D. "Unified Approach to Ultrasonic Light Diffraction." *IEEE Transactions on Sonics and Ultrasonics*, **SU-14** (3) 123–134, July 1967.

4-14. Hance, H. V. and Parks, J. K. "Wide-Band Modulation of a Laser Beam, Using Bragg-Angle Diffraction by Amplitude-Modulated Ultrasonic Waves." *Journal Acoustical Society of America*, **38** (1), 14–23, July 1965.

4-15. Thaler, W. J. "Frequency Modulation of an He–Ne Laser Beam via Ultrasonic Waves in Quartz." *Applied Physics Letters*, **5** (2), 29–31, July 15, 1964.

4-16. Raman, C. V. and Nath, N. S. N. "The Diffraction of Light by High Frequency Sound Waves: Part V Generalized Theory." *Proceedings Indian Academy of Science*, **4**, 222–242, 1936.

4-17. Cummins, H. Z. and Knable, N. "Single Sideband Modulation of Coherent Light by Bragg Reflection from Acoustical Waves." *Proceedings IEEE*, **51** (9), 1246, Sept. 1963.

4-18. Fowles, G. R. *Introduction to Modern Optics*, Holt, Rinehart, and Winston, New York, 1968.

4-19. Yariv, A. *Quantum Electronics*, John Wiley, New York, 1967.

4-20. Kaminow, I. P. and Turner, E. H. "Electrooptic Light Modulators." *Proceedings IEEE*, **54** (10), 1374–1390, Oct. 1966.

4-21. Spencer, E. G., Lenzo, P. V., and Ballman, A. A. "Dielectric Materials for Electrooptic, Elastooptic, and Ultrasonic Device Applications." *Proceedings IEEE*, **55** (12), 2074–2108, Dec. 1967.

4-22. Oldham, W. G. and Bahraman, A. "Electrooptic Junction Modulators." *IEEE Journal of Quantum Electronics*, **QE-3** (7), 278–286, July 1967.

4-23. Mason, W. P. "Optical Properties and the Electro-optic and Photoelastic Effects in Crystals Expressed in Tensor Form." *Bell System Technical Journal*, **29** (2), 161–188, Apr. 1950.

4-24. Billings, B. H. "The Electro-Optic Effect in Uniaxial Crystals of the Type XH_2PO_4." *Journal Optical Society of America*, **39** (10), 797–808, Oct. 1949.

4-25. Blattner, D., Miniter, S., and Sterzer, F. "Cuprous Chloride Light Modulators." *Journal Optical Society of America*, **54** (1), 62–68, Jan. 1964.

4-26. Buhrer, C. F., Bloom, L. R., and Baird, D. H. "Electro-Optic Light Modulation Modulation with Cubic Crystals." *Applied Optics*, **2** (8), 839–846, Aug. 1963.

4-27. Baird, D., Buhrer, C. F., and Conwell, E. M. "Optical Frequency Shifting by Electro-Optic Effect." *Applied Physics Letters*, **1** (2), 46–49, Oct. 1, 1962.

4-28. Macek, W. M., Kroeger, R., and Schneider, J. R. "Microwave Modulation of Light." *IEEE International Convention Record*, **10**, Part 3, 158–176, 1962.

4-29. Jones, O. C. "Methods of Modulating Light at Extreme Frequencies." *Journal Scientific Instruments*, **41**, 653–661, 1964.

4-30. Kaminow, I. P. "Microwave Modulation of the Electro-Optic Effect in KH_2PO_4." *Physical Review Letters*, **6** (10), 528–529, May 15, 1961.

4-31. Blumenthal, R. M. "Design of a Microwave-Frequency Light Modulator." *Proceedings IRE*, **50** (4), 452–456, April 1962.

4-32. Peters, C. J. "Gigacycle Bandwidth Coherent Light Traveling-Wave Phase Modulator." *Proceedings IEEE*, **51** (1), 147–153, Jan. 1963.

4-33. Rigrod, W. W. and Kaminow, I. P. "Wide-Band Microwave Light Modulation." *Proceedings IEEE*, **51** (1), 137–139, Jan. 1963.

4-34. Buhrer, C. F. "Wide-Band Electrooptic Light Modulation Utilizing an Asynchronous Traveling-Wave Interaction." *Applied Optics*, **4** (5), 545–550, May 1965.

4-35. Peters, C. J. et al. "Laser Television System Developed with Off-the-Shelf Equipment." *Electronics*, **38** (3), 75–78, Feb. 8, 1965.

chapter 5

DETECTORS

Optical detectors in a laser communication system convert the intensity of the optical carrier into an electrical signal. For some forms of laser modulation, detectors directly reconstruct the information signal but, in general, detectors are simply components in laser communication receivers. Such communication receivers are discussed in Chapter 10. This chapter contains a summary of detector performance criteria and a discussion of the operating principles of photodetectors. Detection noise is considered in detail in Chapter 8.

Photodetectors fall into two classes: photon and thermal detectors. The latter are of little interest because of their inherently low bandwidth and are not considered further. Photon detectors can be subdivided according to their physical operating mechanism as follows: (a) photoemissive effect, (b) photoconductive effect, (c) photovoltaic effect, and (d) photoelectromagnetic effect.

The photoemissive effect involves the emission of electrons from a vacuum tube cathode in response to optical excitation. The latter three types of photo effects are all associated with semiconductor detectors in which the absorption of photons leads to a change in the concentration of charge carriers in the material.

5.1 PERFORMANCE CRITERIA

The central consideration in the selection of a photodetector for a laser communication optical receiver is the relative output levels of the detector due to a laser carrier and due to noise. Many measures of optical detection performance exist. The following list contains the most widely used performance criteria [5-1, 5-2].

1. Quantum efficiency—η-ratio of the average number of electrons emitted or generated by the detector to the average number of incident photons.

2. Responsivity—$\mathscr{R}_o$-(also called sensitivity) ratio of the average detector current to the average incident radiation power. Responsivity data are

sometimes given in photometric units as amps per lumen. Such data must be corrected for the response of the eye to be useful for communication purposes.

3. Spectral response—variation in detector response as measured by quantum efficiency or responsivity as a function of optical radiation wavelength.

4. Frequency response—when the intensity of the optical radiation incident upon a detector is sinusoidally modulated, the detector output current peak amplitude generally decreases as the modulation frequency is increased. The frequency response is often stated to be the modulation frequency for which the peak output current is one-half its maximum value.

5. Impulse response—when a light pulse strikes a detector the output current generally displays the form shown in Figure 5-1. The time constant τ_0 describes the speed at which the output decays to zero.

6. Dark current, I_D—detector current when the detector is shielded from optical radiation.

7. Noise equivalent power, NEP—rms value of sinusoidally modulated optical radiation which produces an rms signal current equal to the rms detector noise voltage. Statements of the NEP of a detector require knowledge of the modulation frequency, detector bandwidth, detector temperature, and detector area. Typically the reference bandwidth is 1 Hz at a modulation frequency of 400 Hz.

8. Detectivity, D—reciprocal of noise equivalent power. $D \equiv 1/NEP$.

9. D-star, $D^*(\lambda, f)$—detectivity multiplied by the square root of the detector area, A_D, and the square root of the detector bandwidth, Δ_f. $[D^*(\lambda, f) \equiv D\sqrt{A_D \, \Delta_f}]$. Specification of D-star requires knowledge of the modulation frequency and the detector temperature.

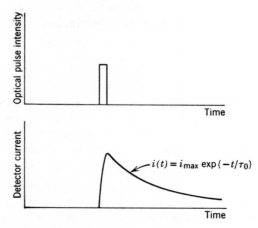

Figure 5-1   *Photodetector impulse response*

The quantum efficiency of a detector is the prime performance characteristic for laser communication applications. Responsivity is of interest only when the absolute magnitude of the detector current is important as, for example, when the detector is directly fed to other circuits without amplification. Performance parameters *NEP*, *D*, and *D** are of limited usefulness for choosing between various types of photodetectors because they provide no inherent information as to the noise source or its spectral content.

In the selection of a photodetector for a laser communication system, the following simple rule can be applied. Among all detectors possessing sufficient bandwidth and acceptable physical characteristics, choose the detector with:

1. the highest quantum efficiency for background radiation limited operation, or

2. the highest *D** (lowest *NEP*) for internal detector noise limited operation.

5.2 PHOTOEMISSIVE DETECTORS

Operation of photoemissive detectors is based upon the external photo-electric effect whereby photons are absorbed by certain solid materials and electrons are emitted [5-3]. Figure 5-2a illustrates the principal elements of a vacuum tube photoemissive detector, commonly called a phototube. A

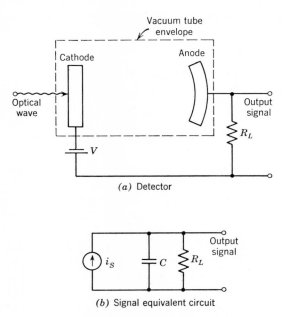

(a) Detector

(b) Signal equivalent circuit

Figure 5-2   *Photoemissive detector*

photoemissive material is coated on the cathode of the phototube, and electrons emitted from the cathode-material are collected by the anode to produce a signal current through the load resistor, R_L.

The maximum energy, ξ_e, of an emitted electron is

$$\xi_e = hf_c - w \tag{5-1}$$

where w is the work function of the material. Only photons above a threshold frequency have energy larger than the work function of the material and, therefore, are able to free electrons. For example, the element cesium has a work function of 1.9 electron volt and its wavelength limit for photoemission is about 0.65 micron. Alkali metals and semiconductor compounds have work functions that permit photoemissive operation for wavelengths of about 1 micron or less.

When an electron within a photoemissive material is excited by optical radiation, it wanders through the material in a random fashion. Energy is lost by the electron with every collision with atoms in the material. Eventually the excited electron may appear at the vacuum surface, and it will be emitted if it has energy greater than the potential barrier at the surface. For a photoemissive material the quantum efficiency, which gives the ratio of the average number of emitted electrons to the average number of incident photons, is a measure of the collision losses and initial photon energy of the excited electrons. Figure 5-3 gives the quantum efficiency of some commercially available photosurfaces as a function of wavelength. The S code numbers refer to the photoemissive materials listed below [5-4, 5-5].

Code	Photoemissive Materials
S-1	Ag-O-Cs
S-10	Ag-Bi-O-Cs
S-17	Cs-Sb
S-20	Sb-K-Na-Cs

Absorption of optical radiation by the glass envelope of the photoemissive tube must be taken into consideration either in the specification of the spectral response curve or the calculation of the receiver transmissivity. In the visible region, glass envelopes for phototubes generally have transmissivities of 90% or greater.

Figure 5-2b shows the signal equivalent circuit of a photoemissive detector represented as a signal current generator paralleled by the detector shunt capacitance. Typically, the shunt capacitance is about 10 to 20 pf. If a modulated laser carrier of instantaneous intensity $C_M(t)$ is incident upon the surface of a photoemissive detector, the instantaneous signal current, i_S, is equal to the instantaneous number of signal photons at the carrier frequency,

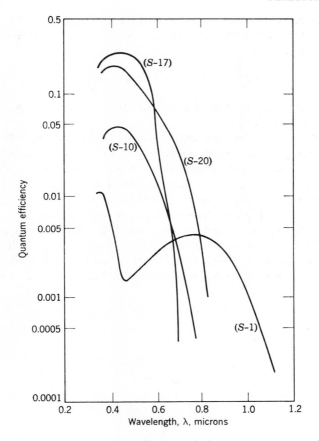

Figure 5-3 *Quantum efficiency of photoemissive materials*

$\widetilde{C_M(t)}/hf_c$,† multiplied by the detector quantum efficiency, η, and the charge of an electron, q. Thus,

$$i_S = \frac{\eta q}{hf_c} \widetilde{C_M(t)} \tag{5-2}$$

This relationship has been found experimentally to hold for frequencies up to 25 GHz [5-6].

In a photoemissive detector a small current is found to flow in the absence of any external photoexcitation. This current, called dark current, is caused primarily by thermal emission, field emission, and current leakage in the detector. The latter two causes can usually be made negligible by proper

† The time average is taken over the carrier period. If the carrier is unmodulated, $\widetilde{C_M(t)} = P_C$.

construction practices and operating conditions. The average dark current due to thermal excitation of electrons in the cathode material obeys Richardson's law [5-7].

$$I_D = 1.2 \times 10^6 A_k T^2 \exp\left\{-\frac{w}{kT}\right\}$$ (5-3)

where A_k is the cathode area, T is the cathode temperature, and w is the cathode material work function.

Figure 5-4 illustrates the dark current temperature variation for two photosensitive surfaces. Dark current emissions not only cause the detector signal to be biased to some positive level but also introduce dark current shot noise. This noise source is discussed further in Chapter 8.

Internal gain is possible with photoemissive detectors by the process of secondary emission. Figure 5-5 shows a photomultiplication chain in which electrons emitted by the cathode successively strike elements called dynodes, each of which is set at a successively larger potential than its predecessor. At each dynode, additional electrons are released, due to secondary emission, and eventually collected at the anode. A gain of 3 to 6 per stage is attainable

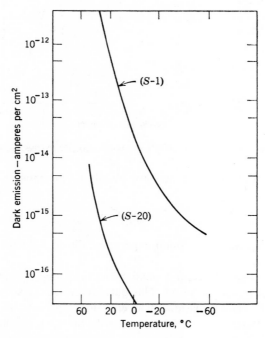

Figure 5-4 *Dark current temperature variation for two photoemissive surfaces* [5-4] (*Courtesy of Radio Corporation of America*)

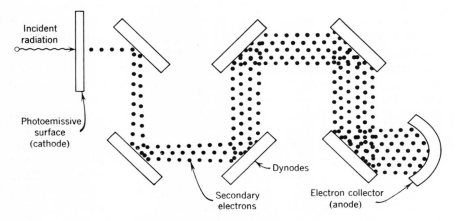

Figure 5-5   *Photomultiplication chain*

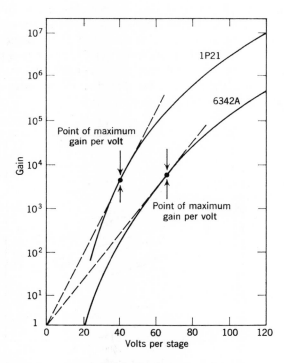

Figure 5-6   *Photomultiplier gain as a function of voltage per stage for two tubes [5-4] (Courtesy of Radio Corporation of America)*

by photomultiplication, yielding a total current gain for a typical nine- or ten-stage photomultiplier tube of about 10^5 to 10^6. The total current gain as a function of the voltage per stage of two commercially available photomultiplier tubes is shown in Figure 5-6 [5-4]. The advantage of gain by the secondary emission process rather than by an electronic amplifier is that with the latter, resistive elements in the input stages of the amplifier cause thermal noise. The thermal noise is amplified along with the photocurrent. In a photomultiplier the photocurrent is amplified before being applied to a load resistor to generate a signal voltage.

In photomultiplier tubes there is a time delay of the order of 50 to 100 nsec from the application of a light pulse to the observation of an anode current [5-4]. The delay is due in small part to the transit time of an excited electron in the photoemissive material and, to a large degree, to the transit time through the dynode chain from cathode to anode. Figure 5-7 illustrates the transit time delay for several photomultiplier tubes.

The finite transit time does not affect the frequency response of the photomultiplier but merely causes a delay in response. Frequency response is limited by the variations in transit time of emitted electrons and by the tube capacitance. Transit time spread is caused by the variation in velocity of primary electrons leaving the photoemissive surface and also secondary electrons leaving each dynode. In addition, secondary electrons emitted from different parts of a dynode often travel different path lengths to the anode, causing a spread in transit time. The transit time spread is on the order of 25 nsec for typical high-quality photomultiplier tubes [5-4].

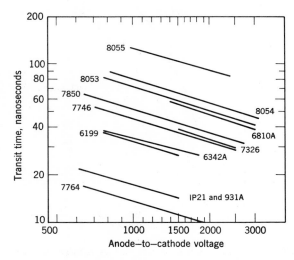

Figure 5-7   *Transit time of several photoemissive tubes* [5-4] (*Courtesy of Radio Corporation of America*)

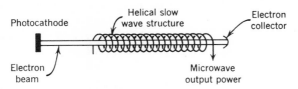

Figure 5-8   *Schematic drawing of a traveling wave phototube*

Several techniques have been proposed for improving the frequency response of phototubes and photomultipliers [5-8]. Phototubes have been operated at frequencies of up to 10 GHz by mounting the tubes in a microwave circuit [5-9]. However, the lack of current gain limits the usefulness of the device. Wide bandwidth and high gain have been obtained by placing the photoemissive elements within a traveling wave structure or crossed-field device. These photodetectors are discussed below.

Figure 5-8 is a schematic drawing of a traveling wave phototube [5-10]. Electrons emitted from the cathode travel along the axis of a helical slow wave structure to an electron collector. Modulation of the electron beam is extracted by the helix to an output circuit. At low electron beam currents the traveling wave phototube does not produce a current gain, but the equivalent load resistance of the tube can be made quite large without sacrificing frequency response [5-11]. An equivalent load resistance greater than 10^5 ohms has been achieved in a frequency band of 10 to 20 GHz with a traveling wave phototube [5-12]. Despite the relatively large load resistance, the traveling wave phototube is generally thermal noise limited.

The transit time spread of the photomultiplication chain has been reduced considerably in the static crossed-field photomultiplier by the simple expedient of reducing transit time from dynode to dynode [5-13]. Figure 5-9 illustrates the major elements of the static crossed-field photomultiplier. Electrons emitted from the photocathode or any dynode are accelerated by the large electric field between the cathode or dynode and the dummy electrode. The magnetic field causes these emitted electrons to follow a

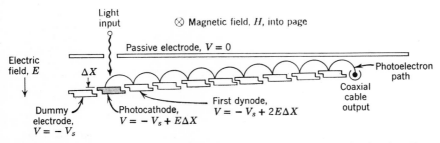

Figure 5-9   *Schematic drawing of static crossed-field photomultiplier [5-13]*

circular path from dynode to dynode. High gain at frequencies up to 5 GHz has been reported for the tube [5-13].

The dynamic crossed-field electron multiplier (DCFEM) is another device that makes use of combined electric and magnetic fields to reduce transit time dispersion [5-14]. Figure 5-10 contains a sketch of a DCFEM tube. A microwave frequency electric field accelerates electrons emitted from the photocathode dynode during a portion of each microwave field cycle. The magnetic field causes the emitted electrons to follow a curved trajectory to the active electrode. Electrons emitted by secondary emission from the active electrode are likewise accelerated by the microwave field. The resultant output is a series of current pulses at a repetition rate equal to the microwave field frequency. These current pulses are essentially samples of the modulation signal. Hence, by the sampling theorem, the theoretical baseband modulation limit of the detector is one-half the microwave pump frequency.

5.3 PHOTOCONDUCTIVE DETECTORS

In a photoconductive detector, incident photons cause an additional concentration of charge carriers, thereby allowing a larger current from an external circuit to flow through the detector [5-15, 5-16]. Essentially, the resistance of the photoconductor decreases with an increase in the amount of incident optical radiation. Figure 5-11 illustrates the static current-voltage characteristics of a photoconductive detector.

Photoconductive materials are generally classed as semiconductors or insulators. The distinction is based upon the magnitude of the energy gap between the valance and conduction bands of the material. Insulators have energy gaps greater than 2 electron volts; semiconductors have energy gaps somewhat less than that amount. Semiconductor photoconductive materials

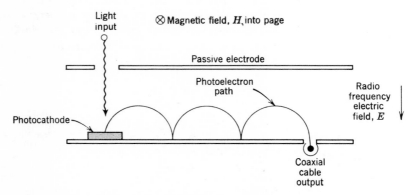

Figure 5-10 *Schematic drawing of dynamic crossed-field photomultiplier*

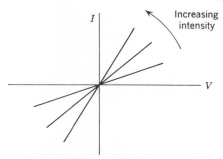

Figure 5-11   *Current-voltage characteristics of a photoconductive detector*

are further classified as being intrinsic or extrinsic. Figure 5-12 contains energy level diagrams of intrinsic and extrinsic semiconductors. The intrinsic semiconductor is characterized by valence and conduction bands separated by an energy gap or forbidden band with no intervening energy levels. In the intrinsic semiconductor, optical radiation causes electrons to leave the valence band and enter the conduction band. The absence of the bound electron in the valence band is called a hole. Holes are free to move within the valence band. Current flow is obtained by the propagation of electrons in the conduction band and by the movement of holes in the valence band. In the extrinsic semiconductor the semiconductor material is not pure but contains imperfections such as impurities, dislocations, and lattice defects. These imperfections create additional energy levels which may occur within the forbidden gap. Without external excitation, electrons and holes in the forbidden gap remain fixed since normal conduction does not occur in the forbidden band. However, optical radiation may cause electrons in the forbidden gap to move into the conduction band and electrons in the valence band to move into the forbidden band, leaving a hole in the valence band. This process adds to the conductivity of the material.

For a photon to be absorbed by an intrinsic semiconductor, it is necessary that the photon energy be greater than the energy gap. This places a long wavelength limit, λ_c, on the photoconduction process given by

$$\lambda_c = \frac{hc}{E_g} \tag{5-4}$$

where E_g is the energy gap energy. The long wavelength limit is not sharp due to temperature variations in E_g and other more complex factors. The long wavelength limit for extrinsic semiconductors is somewhat more complicated because of the additional energy levels in the forbidden gap. A short wavelength limit on the spectral response is set by the surface absorption of optical radiation.

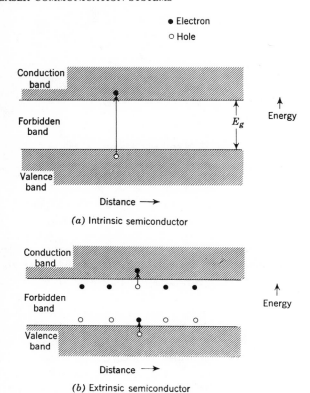

Figure 5-12   Energy level diagrams of intrinsic and extrinsic semiconductors

The following is a list of some commonly employed photoconductive materials.

Intrinsic Semiconductors	
PbS	Lead sulphide
PbSe	Lead selenide
PbTe	Lead tellurium
InSb	Indium antimonide

Extrinsic semiconductors	
Ge	Doped germanium
Si	Doped silicon

Insulators	
CdS	Cadmium sulphide
CdSe	Cadmium selenide

The quantum efficiency of photoconductive detectors, within the frequency range established by the long and short wavelength limits, is relatively high, approaching unity for some detectors [5-17, 5-18]. Unfortunately, quantum efficiency is usually not specified by manufacturers of photoconductive detectors but rather the noise equivalent power, *NEP*, or detectivity, D^*, of the detector is given. In most measurements of *NEP* and D^*, the limiting noise is current or $1/f$ noise at low frequencies and generation-recombination noise at high frequencies [5-19]. Typical D^* measurements of intrinsic, extrinsic, and insulator photoconductors are presented in Figures 5-13 and 5-14 [5-20].

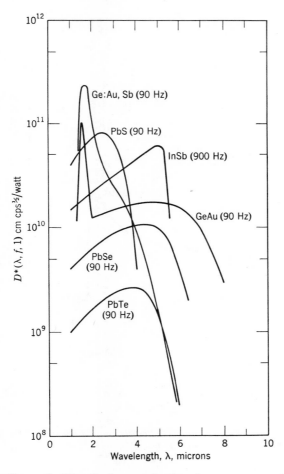

Figure 5-13 *Spectral D^* of photoconductive detectors operating at infrared frequencies, detector temperature 77° K [5-20]*

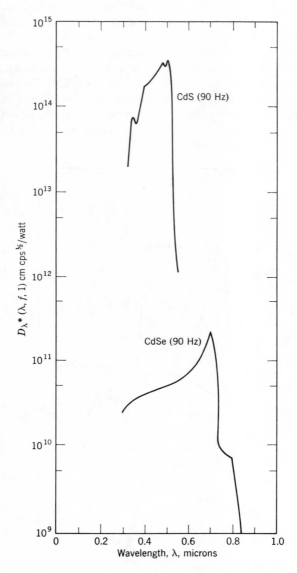

Figure 5-14   *Spectral D* of photoconductive detectors operating at visible frequencies, detector temperature 295° K [5-20]*

Figure 5-15a is a schematic drawing of a photoconductive detector with a direct current bias. Radio frequency biasing will be considered later. The signal equivalent circuit of the photoconductive detector shown in Figure 5-15b includes a signal current source in parallel with the shunt capacitance and equivalent resistance of the detector. The relationship between the signal current and optical radiation power is not independent of the modulation frequency of the radiation source as for a photoemissive detector. If the laser carrier incident on the detector is intensity modulated by a sinusoid of angular frequency ω_m, with a 100% intensity modulation index, the instantaneous carrier intensity may be expressed as

$$C_M(t) = \frac{A_c^{\,2}}{2} \left[1 + \cos \omega_m t\right] \cos^2 \omega_c t \qquad (5\text{-}5)$$

The corresponding number of signal photons incident upon the detector per second is

$$\frac{\widetilde{C_M(t)}}{hf_c} = \frac{A_c^{\,2}}{4} \left[1 + \cos \omega_m t\right] \equiv \frac{P_C}{2} \left[1 + \cos \omega_m t\right] \qquad (5\text{-}6)$$

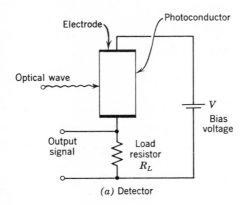

(a) Detector

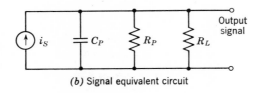

(b) Signal equivalent circuit

Figure 5-15 *Photoconductive detector with direct current bias*

where P_C is the average unmodulated carrier power. The photoconductive detector output signal current is then given by

$$i_S = \frac{\eta q}{hf_c} \frac{\tau_L}{\tau_T} \frac{1}{(1 + \omega_m^2 \tau_L^2)^{1/2}} \frac{P_C}{2} [1 + \cos(\omega_m t + \Upsilon)] \qquad (5\text{-}7)$$

where Υ is a phase shift of the signal current which is dependent upon ω_m, τ_L is the average lifetime of an electron in the conduction band, and τ_T is the transit time for an electron to move through the detector [5-17]. The ratio

$$G_{PC} \equiv \frac{\tau_L}{\tau_T} \qquad (5\text{-}8)$$

is called the gain of a photoconductor. Photoconductor gains can be larger or smaller than unity depending upon the ratio of τ_L to τ_T [5-21].

The carrier transit time is given by

$$\tau_T = \frac{L_E^2}{V_o \mu_e} \qquad (5\text{-}9)$$

where L_E is the spacing between electrodes, V_o is the direct current bias voltage, and μ_e is the electron mobility [5-16]. Substitution of Equation 5-9 into Equation 5-7 shows that the photocurrent increases linearly with applied voltage. The increase in signal current is limited, however, by space charge. When space charge conditions are reached the transit time is limited by the electrode capacitance, and becomes equal to the dielectric relaxation time, τ_D. The maximum photoconductor gain is then

$$[G_{PC}]_{\max} = \frac{\tau_L}{\tau_D} \qquad (5\text{-}10)$$

Figures 5-16 and 5-17 give the measured frequency response for several

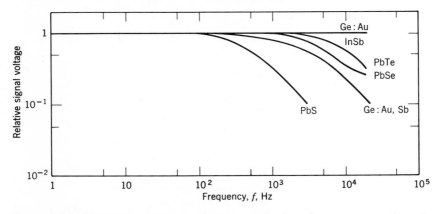

Figure 5-16 *Frequency response of photoconductive detectors operating at a detector temperature of 77° K [5-20]*

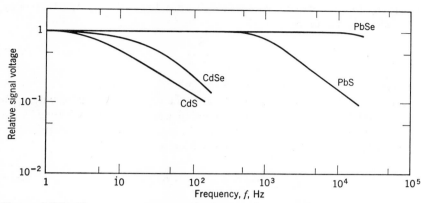

Figure 5-17 *Frequency response of photoconductive detectors operating at a detector temperature of 295° K [5-20]*

photoconductors. In a photoconductive detector the frequency response, measured by the bandwidth, B, is limited by the carrier lifetime, τ_L.

$$B < \frac{1}{2\pi\tau_L} \qquad (5\text{-}11)$$

By combining Equations 5-10 and 5-11, it is seen that the photoconductor exhibits a gain-bandwidth product restriction which is equal to

$$G_{PC}B < \frac{1}{2\pi\tau_D} \qquad (5\text{-}12)$$

The gain-bandwidth product for a photoconductive detector with direct current bias is a result of the ohmic contacts which inject a space charge when the bias voltage is sufficiently high. If a radio frequency (RF) bias is used, as shown in Figure 5-18, capacity coupling may be employed and, hence,

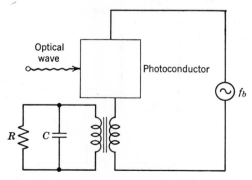

Figure 5-18 *Photoconductive detector with radio frequency bias*

conducting contacts will not be required [5-15, 5-16]. If the RF bias frequency, f_b, satisfies the inequality

$$f_b > \frac{1}{2\pi\tau_T} \tag{5-13}$$

the electric field will reverse before the minority carriers leave the photoconductor. Under this condition the hole-electron pairs oscillate within the material and the carrier lifetime is then independent of the transit time [5-15]. Then, from Equations 5-8, 5-11, and 5-13 the gain-bandwidth product for a photoconductor with RF bias may be written as

$$G_{PC}B < \frac{1}{2\pi\tau_T} < f_b \tag{5-14}$$

Theoretically, the gain of a photoconductor can be increased to any level desired for a fixed bandwidth by simply increasing the bias frequency. In practice, the driving frequency must be 10 to 100 GHz to achieve the same gain as a photomultiplier for bandwidths of only 100 kHz.

5.4 PHOTOVOLTAIC AND PHOTOELECTROMAGNETIC DETECTORS

Photovoltaic Detectors

In a semiconductor photovoltaic detector (Figure 5-19), photons, which are absorbed in the region of a semiconductor *p-n* junction, produce an output potential across the junction caused by the diffusion of hole-electron pairs [5-22]. The voltage is produced without a bias supply or load resistor.

Figure 5-20 shows the energy level diagram of a *p-n* semiconductor junction. The Fermi levels in the *p-* and *n*-regions are aligned in the absence of optical

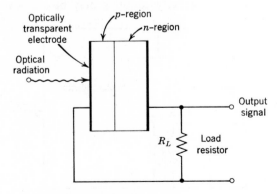

Figure 5-19   Photovoltaic detector

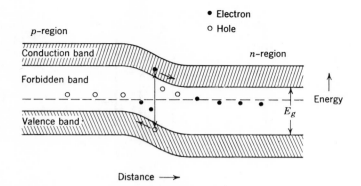

Figure 5-20 *Energy level diagram of a semiconductor p-n junction*

radiation, and an internal electric field exists at the junction. When a photon of the proper wavelength is absorbed in the region of the junction, a free hole-electron pair is produced. The internal electric field splits the pair, moving the hole into the *p* material and the electron into the *n* material. This movement of electrons and holes causes the *n* material to be charged negatively and the *p* material to be charged positively. An open circuit voltage will then be generated across the detector, and if a load resistor is connected across the device, a current will flow.

Frequency response in a photovoltaic detector is limited by the drift velocity of free carriers in the junction region. By placing an electric field across the junction, the velocity of free carriers can be made equal to the saturated drift velocity. If a direct current bias voltage is applied to a photovoltaic junction, as in Figure 5-21*a*, the device is called a photodiode [5-23 to 5-25]. Figure 5-22 illustrates the current-voltage characteristics of a photodiode as a function of light level. Without any optical radiation, the device acts as an ordinary diode. Light falling on the junction produces a linear increase in current for reversed bias operation. The increase in current is detected by a load resistor, as with a photoconductive detector. Note, however, the difference in current-voltage characteristics between a photodiode and photoconductor.

Some types of photodiodes contain a region of intrinsic material sandwiched between the *p*- and *n*-regions [5-23]. The reason for this configuration is to increase the junction width and thus decrease the junction capacitance.

The wavelength response of a photodiode is governed by the same factors as a photoconductor. Optical radiation must have photon energy greater than the energy gap, E_g, of the material in order to produce free carriers. This establishes the long wavelength limit. Short wavelength radiation is

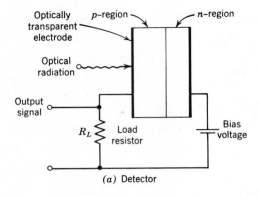

(a) Detector

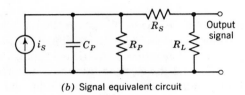

(b) Signal equivalent circuit

Figure 5-21   Photodiode detector

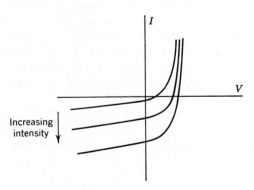

Figure 5-22   Current-voltage characteristics of a photodiode detector

absorbed at the surface of the detector. Figure 5-23 is a plot of the quantum efficiency of typical germanium and silicon photodiodes [5-26].

The signal equivalent circuit of a photodiode detector is shown in Figure 5-21b. In addition to shunt capacitance and shunt resistance, the circuit includes a series resistance which is typically 1 to 100 Ω. The signal current of a photodiode is given by

$$i_S = \frac{\eta q}{h f_c} \widetilde{C_M(t)} \tag{5-15}$$

This equation applies to modulation frequencies to about 10 GHz.

In a photodiode the frequency response is dependent upon the diffusion time of carriers to the junction, drift time in the junction, and junction capacitance. The diffusion time can be reduced by constructing the diode with its junction close to the surface of incident radiation. Then the optical radiation is absorbed in the junction region. The capacitance, C_P, of the junction region is

$$C_P = \frac{\epsilon A_J}{W_J} \tag{5-16}$$

where ϵ is the dielectric constant of the semiconductor material, A_J is the junction area, and W_J is the junction thickness. From the signal equivalent of Figure 5-21b, if the load resistor is small, as it must be for high frequency operation, then the cutoff frequency, ω_q, becomes

$$\omega_q = \frac{(1 + R_S/R_P)}{R_S C_P} \approx \frac{1}{R_S C_P} \tag{5-17}$$

The approximation is valid when the series resistance is much smaller than

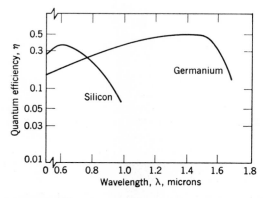

Figure 5-23   *Quantum efficiency of silicon and germanium photodiodes* [5-26]

the shunt resistance as is the case for all practical photodiodes. Hence, to increase the $R_S C_P$ cutoff frequency of the photodiode, it is necessary to make R_S and C_P as small as possible. To reduce C_P, the junction area of fast photodiodes is generally held to 10^{-4} cm^2 to 10^{-3} cm^2 [5-25]. This, however, tends to create problems in focusing the optical radiation to the detector surface. The junction width, W_J, cannot be increased greatly to reduce C_P since the carrier transit time in the junction is proportional to W_J. With proper design, junction capacitance can be reduced sufficiently so that the junction drift time is the limiting factor on frequency respose.

Internal gain has been achieved in a photodiode by the avalanche electron process [5-27, 5-28]. Free electrons in the diode are subject to a high reverse bias field which causes ionizing collisions with atoms. Each collision produces a secondary electron which travels along with the primary electron and a secondary hole which travels in the opposite direction. The secondary hole also has the potential of creating another hole-electron pair by an ionizing collision. With the avalanche process, gains of up to 250 have been obtained for relatively high modulation frequencies. There is an excess shot noise produced by the gain mechanism, however [5-8].

Photoelectromagnetic Detectors

If a photoconductive material is subjected to a magnetic field orthogonal to the propagation direction of incident radiation a current will be produced due to the photoelectromagnetic effect [5-22, 5-29]. The photoelectromagnetic effect is related to the Hall effect. Optical radiation incident upon the detector surface creates hole-electron pairs which diffuse downward from the illuminated surface. The magnetic field separates each pair by the Hall angle. Splitting of the carriers produces an internal field which causes a voltage to be generated across the material. The instantaneous signal current follows the relationship of Equation 5-2 for frequencies up to 10 GHz [5-22].

REFERENCES

5-1. Jones, R. C. "Performance of Detectors for Visible and Infrared Radiation." *Advances in Electronics, Vol. V.* L. Marton, ed., Academic Press, 1953.

5-2. Jones, R. C. "Quantum Efficiency of Detectors for Visible and Infrared Radiation." *Advances in Electronics, Vol. XI.* L. Marton, ed., Academic Press, 1959.

5-3. Spicer, W. E. and Wooten, F. "Photoemission and Photomultipliers." *Proceedings IEEE,* **51** (8), 1119–1126, Aug. 1963.

5-4. "RCA Phototubes and Photocells." Radio Corporation of America, Technical Manual PT-60, 1963.

5-5. "Dumont Multiplier Phototubes." Dumont Electron Tubes Manual, June 1965.

5-6. Johnson, A. M. "Square Law Behavior of Photocathodes at High Light Intensities and High Frequencies." *IEEE Journal of Quantum Electronics*, **QE-1**, 99–101, May 1965.

5-7. Lallemand, A. *Astronomical Techniques*. W. Hiltner, ed. Univ. of Chicago Press, 1962.

5-8. Anderson, L. K. and McMurtry, B. J., "High-Speed Photodetectors." *Proceedings IEEE*, **54** (10), 1335–1349, Oct. 1966.

5-9. Lindsay, P. A. et al. "Optical Mixing in Phototubes." *Proceedings IEEE Correspondence*, **50** (11), 2380–2381, Nov. 1962.

5-10. McMurtry, B. J. and Siegman, A. E. "Photomixing Experiments with a Ruby Optical Maser and a Traveling Wave Microwave Phototube." *Applied Optics*, **1**, 51–53, Jan. 1962.

5-11. McMurtry, B. J. "Microwave Phototube Design Considerations." *IEEE Transactions on Electron Devices*, **ED-10**, 219–226, July 1963.

5-12. Caddes, D. E. "A Ku-Band Traveling Wave Phototube." *Microwave Journal*, **8**, 3–8, Mar. 1965.

5-13. Miller, R. C. and Wittwer, N. C. "Secondary-Emission Amplification at Microwave Frequencies." *IEEE Journal of Quantum Electronics*, **QE-1**, 49–59, April 1965.

5-14. Gaddy, O. L. and Holshouser, D. F. "A Microwave Frequency Dynamic Crossed-Field Photomultiplier." *Proceedings IEEE*, **51**, 153–162, Jan. 1963.

5-15. Sommers, H. S., Jr., and Teutsch, W. B. "Demodulation of Low-Level Broad-Band Optical Signals with Semiconductors, Part II: Analysis of the Photoconductive Detector." *Proceedings IEEE*, **52** (2), 144–153, Feb. 1964.

5-16. Sommers, H. S. Jr., and Gatchell, E. K. "Demodulation of Low-Level Broad-Band Optical Signals with Semiconductors, Part III: Experimental Study of the Photoconductive Detector." *Proceedings IEEE*, **54** (11), 1553–1568, Nov. 1966.

5-17. DiDomenico, M., Jr., and Anderson, L. K. "Microwave Signal-to-Noise Ratio Performance of CdSe Bulk Photoconductive Detectors." *Proceedings IEEE*, **52** (7), 815–822, July 1964.

5-18. DiDomenico, M., Jr., and Svelto, O. "Solid-State Photodetection: A Comparison Between Photodiodes and Photoconductors." *Proceedings IEEE*, **52** (2), 136–144, Feb. 1964.

5-19. Wolfe, W. L. ed. *Handbook of Military Infrared Technology*. Office of Naval Research, U.S. Government Printing Office, Washington, 1965.

5-20. Kruse, P. W., McGlauchlin, L. D., and McQuistan, R. B. *Elements of Infrared Technology: Generation, Transmission, and Detection*. John Wiley, New York, 1962.

5-21. Ing, S. W., Jr., and Gerhard, G. C. "A High Gain Silicon Photodetector." *Proceedings IEEE*, **53** (11), 1714–1722, Nov. 1965.

5-22. Sommers, H. S. Jr. "Demodulation of Low-Level Broad-Band Optical Signals with Semiconductors. "*Proceedings IEEE*, **51** (1), 140–146, Jan. 1963.

5-23. Riesz, R. P. "High Speed Semiconductor Photodiodes." *Review of Scientific Instruments*, **33** (9), 994–998, Sept. 1962.

5-24. Lucovsky, G., Lasser, M. E., and Emmons, R. B. "Coherent Light Detection in Solid-State Photodiodes." *Proceedings IEEE*, **51** (1), 166–172, Jan. 1963.

5-25. Lucovsky, G. and Emmons, R. B. "High Frequency Photodiodes." *Applied Optics*, **4** (6), 697–702, June 1965.

5-26. Melchior, H. and Lynch, W. T. "Signal and Noise Response of High Speed Germanium Avalanche Diodes." *IEEE Transactions on Electron Devices*, **ED-13** (12) 829–838, Dec. 1966.

5-27. Lasser, M. E. "Detection of Coherent Optical Radiation." *IEEE Spectrum*, **3** (4), 73–78, April 1966.

5-28. Johnson, K. M. "High-Speed Photodiode Signal Enhancement at Avalanche Breakdown Voltage." *IEEE Transactions on Electron Devices*, **ED-12**, 55–63, Feb. 1965.

5-29. van Roosebroeck, W. "Theory of the Photomagnetoelectric Effect in Semiconductors." *Physical Review*, **101**, 1713–1725, Mar. 15, 1956.

chapter 6

BACKGROUND RADIATION

Optical receiver performance is often limited by background radiation from the sun, moon, planets, stars, and sky. Background radiation impairs laser signal detection by increasing the detector shot noise level.

The following sections describe the most widely used measures of background radiation. Typical data available in the literature are given for various background radiation sources.

6.1 BLACKBODY MODEL OF BACKGROUND RADIATION

Background radiation results from sources, which are at an elevated temperature, producing self-emissions and from sources that reflect radiation from other hot bodies. Radiation from such sources often can be modeled by Planck's law of blackbody radiation [6-1]. One characterization of Planck's law is the spectral radiant emittance of a body in terms of the power emitted per unit area of the source into a hemisphere in the incremental wavelength region λ to $\lambda + d\lambda$. In wavelength units, for unpolarized radiation, Planck's law takes the form

$$\mathscr{W}(\lambda)\, d\lambda = \frac{2\pi c^2 h}{\lambda^5} \frac{1}{[\exp\{hc/\lambda k T_S\} - 1]}\, d\lambda \tag{6-1}$$

where:

$\mathscr{W}(\lambda)$ = background radiation, spectral radiant emittance in wavelength units
c = velocity of light
λ = wavelength
T_S = temperature of background radiation source
h = Planck's constant
k = Boltzmann's constant

The spectral radiant emittance can also be expressed in frequency units by the relations

$$\mathscr{W}(\lambda)\, d\lambda = -\mathscr{W}(f)\, df \quad \text{and} \quad df = -\frac{c}{\lambda^2}\, d\lambda \tag{6-2}$$

111

yielding

$$\mathscr{W}(f)\, df = \frac{2\pi h f^3}{c^2} \frac{1}{[\exp\{hf/kT_S\} - 1]}\, df \qquad (6\text{-}3)$$

where $\mathscr{W}(f)$ is the spectral radiant emittance in frequency units and f is the frequency.

Figure 6-1 illustrates the spectral radiant emittance of a blackbody as a function of wavelength and source temperature. The wavelength maximum for each source temperature, obtained by differentiation of Equation 6-1, is Wein's displacement law

$$\lambda_m = \frac{2.89 \times 10^{-3}}{T_S} \qquad (6\text{-}4)$$

where the wavelength is in meters and the temperature is in degrees Kelvin. Thus, it is possible to characterize radiation from a blackbody source by a single absolute temperature of the source. The integral of the blackbody curve over all wavelengths gives the total spectral energy of the blackbody source radiated into a hemisphere. By the Stefan-Boltzmann law

$$\int_0^\infty \mathscr{W}(\lambda)\, d\lambda = \sigma T_S^4 \qquad (6\text{-}5)$$

where $\sigma = 5.67 \times 10^{-12}$ watt/cm^2 $^\circ$K^4 is the Stefan-Boltzmann constant.

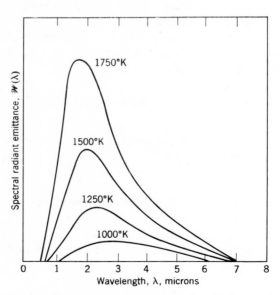

Figure 6-1   *Blackbody spectral radiant emittance*

In an optical communication system, an optical filter is placed at the entrance to the receiver to limit the amount of background radiation entering the receiver. The spectral width of these filters, though wide by electrical standards, is narrow with respect to the range of the blackbody curve. Hence the background radiation intensity usually may be regarded as constant over the optical input filter passband.

The bandwidth of optical filters is usually specified in wavelength units since the background radiation measurements are most often in terms of wavelength. Conversion to frequency units may be made by the relationship

$$B_i = \frac{c}{\lambda_c^2} \lambda_i \tag{6-6}$$

where B_i is the optical filter bandwidth in frequency units and λ_i is the optical filter bandwidth in wavelength units. Figure 6-2 gives the optical filter bandwidth in frequency units for a 1 Å optical filter as a function of the center wavelength of the filter.

6.2 BACKGROUND RADIATION MEASUREMENT

Several radiometric quantities of background radiation measurement have been developed in addition to the spectral radiant emittance [6-2 to 6-5]. Table 6-1 lists some of the most useful characterizations for laser communication system design. The relationships between these characterizations are developed below.

Symbol	Title	Description	Typical Units
$\mathscr{W}(\lambda)$	Spectral radiant emittance	Radiant power into a hemisphere per unit area of source in hemisphere	$\dfrac{\text{watts}}{\text{cm}^2 \text{ micron}}$
$\mathscr{N}(\lambda)$	Spectral radiance	Radiant power into a unit solid angle per unit projected area of source in hemisphere	$\dfrac{\text{watts}}{\text{cm}^2 \text{ micron steradian}}$
$\mathscr{Q}(\lambda)$	Photon spectral radiance	Number of photons per second radiated into a unit solid angle per unit projected area of source in hemisphere	$\dfrac{\text{watts}}{\text{sec cm}^2 \text{ micron steradian}}$
$\mathscr{H}(\lambda)$	Spectral irradiance	Radiant power incident upon a surface per unit surface area	$\dfrac{\text{watts}}{\text{cm}^2}$

Table 6-1. BACKGROUND RADIATION MEASUREMENT QUANTITIES

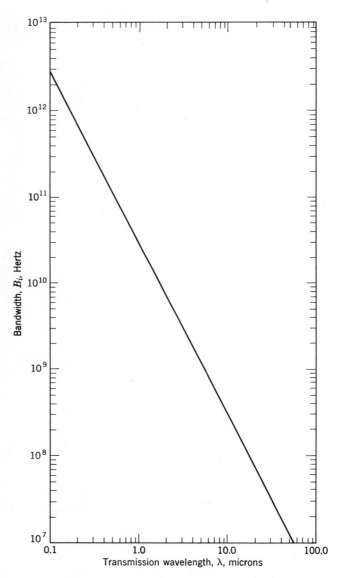

Figure 6-2   *Bandwidth in Hertz of a 1Å optical bandpass filter*

Consider an elemental area, dA, which radiates power into an elemental solid angle, $d\Omega$, as shown in Figure 6-3. The *spectral radiance*, $\mathcal{N}(\lambda)$, of this source, as defined in Table 6-1, is the power per unit wavelength interval radiated into $d\Omega$ from the projected area of the source $dA \cos \theta$. Also, from the table, the spectral radiant emittance, $\mathcal{W}(\lambda)$, is the power per unit wavelength interval radiated into the entire hemisphere from the actual area, dA. Hence, the quantities are related by

$$\mathcal{W}(\lambda) = \int_{\substack{\text{over} \\ \text{hemisphere}}} \mathcal{N}(\lambda) \cos \theta \, d\Omega \tag{6-7}$$

From Figure 6-3 the elemental solid angle is

$$d\Omega = \frac{(R \sin \theta \, d\theta)(R \, d\Phi)}{R^2} = \sin \theta \, d\Phi \, d\theta$$

Thus,

$$\mathcal{W}(\lambda) = \int_{\Phi=0}^{2\pi} \int_{\theta=0}^{\pi/2} \mathcal{N}(\lambda) \cos \theta \sin \theta \, d\theta \, d\Phi \tag{6-8}$$

In general, $\mathcal{N}(\lambda)$ is a function of θ and Φ. However, in many practical

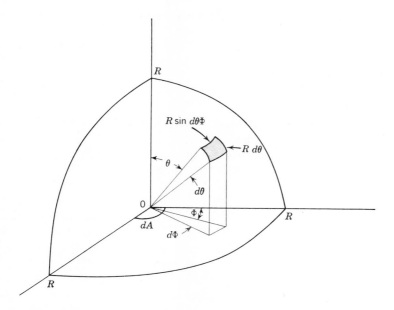

Figure 6-3 *Geometry for relationship between radiant emittance and radiance*

situations the spectral radiance is the same in all directions. Surfaces for which this is true are called Lambertian. The sun and many planets appear as uniformly bright Lambert surfaces. Assuming $\mathscr{N}(\lambda)$ constant, Equation 6-8 integrates directly to yield

$$\mathscr{W}(\lambda) = \pi \mathscr{N}(\lambda) \qquad (6\text{-}9)$$

Hence, the spectral radiant emittance and spectral radiance measurements are related by a constant for Lambertian surfaces.

Spectral radiance measurements are often expressed in terms of the average number of photons incident upon the receiver detector surface. The photon spectral radiance is defined as

$$\mathscr{Q}(\lambda) \equiv \frac{\mathscr{N}(\lambda)}{hf_c} \qquad (6\text{-}10)$$

where f_c is the center frequency of the optical filter.

The spectral radiant emittance and spectral radiance characterizations of background radiation are useful if the actual or projected area of the source is known. Such information is not generally available for stars and other objects which appear as point sources to the receiver. For point sources, background radiation measurements are often made in terms of the *spectral irradiance*, $\mathscr{H}(\lambda)$, of the source. The spectral irradiance is defined to be the power per unit wavelength interval incident upon a unit area of the receiver plane. Measurements of $\mathscr{H}(\lambda)$ are sometimes referenced to the earth's surface and implicitly contain the atmospheric transmissivity factor. For extended sources of circular cross section, the spectral irradiance is related to the spectral radiance by

$$\mathscr{H}(\lambda) = \Omega_S \mathscr{N}(\lambda) \qquad (6\text{-}11)$$

where Ω_S is the solid angle subtended by the source at the receiver.

Given some background radiation measurement, the background radiation power incident upon a receiver detector can be expressed in terms of the measurement and the physical characteristics of the receiver. Several practical cases will now be considered.

Figure 6-4 illustrates the measurement of background radiation power utilizing the spectral radiant emittance of a spherical source, such as a planet, of diameter, d_S, at a distant range, R, from a communication receiver. The total amount of background radiation power per unit wavelength emitted from the source into the hemisphere defined by R is the product of the spectral radiant emittance of the source and the source area within the hemisphere, $(\pi d_S^2 / 2)$. The source is assumed to radiate uniformly in all radial directions—a Lambert source. Then the total background radiation power incident upon the receiver detector, P_B, is equal to the total radiated power

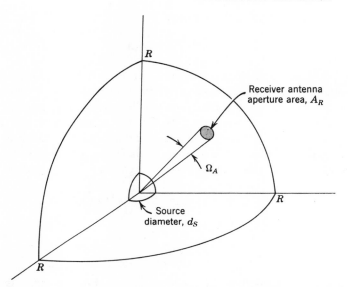

Figure 6-4 *Geometry for spectral radiant emittance background radiation measurements*

passing through the input filter, multiplied by the atmospheric and receiver transmissivities and by the ratio of the receiver area to the area of the hemisphere. Thus

$$P_B = \frac{\tau_a \tau_r \lambda_i d_S{}^2 A_R}{4R^2} \mathscr{W}(\lambda) \qquad (6\text{-}12)$$

where:

P_B = background radiation average power at detector surface
τ_a = atmospheric transmissivity
τ_r = receiver transmissivity
λ_i = input filter bandwidth in wavelength units
R = range
A_R = receiver antenna aperture area
d_S = diameter of background radiation source

Equivalently, P_B is equal to the product of $\tau_a \tau_r \lambda_i$; the spectral radiance, $\mathscr{N}(\lambda)$; the projected circular area of the source, $\pi d_S{}^2/4$ and the solid angle subtended at the source by the receiver antenna, Ω_A. For a circular receiver antenna

$$\Omega_A = \frac{A_R}{R^2} \qquad (6\text{-}13)$$

and the total background radiation power is

$$P_B = \frac{\pi \tau_a \tau_r \lambda_i d_S{}^2 A_R}{4R^2} \mathcal{N}(\lambda) \qquad (6\text{-}14)$$

Equations 6-12 and 6-14 differ only by the factor π which, from Equation 6-9, is the relation between the spectral radiant emittance and the spectral radiance for a Lambertian surface. It has been implicitly assumed in the formulation of equations 6-12 and 6-14 that the entire source is within the receiver field of view. If the solid angle subtended by the source at the receiver, Ω_S, is greater than the solid angle of the receiver field of view, Ω_R, the background radiation power, to a good approximation, is given by

$$P_B = \frac{\tau_a \tau_r \lambda_i d_S{}^2 A_R \Omega_R}{4R^2 \Omega_S} \mathcal{W}(\lambda) \qquad (6\text{-}15)$$

and

$$P_B = \frac{\pi \tau_a \tau_r \lambda_i d_S{}^2 A_R \Omega_R}{4R^2 \Omega_S} \mathcal{N}(\lambda) \qquad (6\text{-}16)$$

The solid angle subtended by the spherical source of diameter, d_S, is

$$\Omega_S = \frac{\pi d_S{}^2}{4R^2} \qquad (6\text{-}17)$$

And for small angles the solid angle receiver field of view is

$$\Omega_R = \frac{\pi \theta_R{}^2}{4} \qquad (6\text{-}18)$$

where θ_R is the planar angle of the receiver field of view. Hence, for a source entirely filling the receiver field of view, the background radiation power reduces to

$$P_B = \frac{\tau_a \tau_r \lambda_i \, \theta_R{}^2 A_R}{4} \mathcal{W}(\lambda) \qquad (6\text{-}19)$$

or

$$P_B = \frac{\pi \tau_a \tau_r \lambda_i \, \theta_R{}^2 A_R}{4} \mathcal{N}(\lambda) \qquad (6\text{-}20)$$

Equations 6-19 and 6-20 are useful formulations for extended sources such as the sky background.

The receiver field of view and receiver antenna size are independent only in that the field of view can be made larger than the diffraction limit. For a circular receiver antenna of diameter d_R, the diffraction limit is

$$\theta_R \geq \frac{\lambda_c}{d_R} \qquad (6\text{-}21)$$

The receiver antenna of a radio frequency communication system is often diffraction limited, but this is seldom the case for optical communications because of difficulties in pointing the receiver antenna if θ_R becomes very small.

If a background radiation measurement is given in terms of the spectral irradiance, $\mathcal{H}(\lambda)$, the background radiation power received is simply the product of $\tau_a \tau_r \lambda_i$, the receiver antenna area, and $\mathcal{H}(\lambda)$. Thus

$$P_B = \tau_a \tau_r \lambda_i A_R \mathcal{H}(\lambda) \qquad (6\text{-}22)$$

Table 6-2 lists the relations for the background radiation power at the surface of a receiver detector for a circular receiver antenna of diameter d_R. The incident power, as evidenced by the equations of Table 6-2, is dependent upon the receiver aperture diameter, the receiver field of view, the size of the source, and the range to the source. For those sources that completely fill the receiver field of view, the spectral radiance, $\mathcal{N}(\lambda)$ or $\mathcal{Q}(\lambda)$, is convenient for describing the background intensity. Stars and distant planets observed through a relatively wide field of view appear as point sources. For these sources the spectral irradiance, $\mathcal{H}(\lambda)$, best characterizes the background

Source Relationship	Expression	Background Radiation Quantity
Any source	$P_B = \dfrac{\pi\ \tau_a \tau_r \lambda_i\ d_R{}^2}{4} \mathcal{H}(\lambda)$	Spectral irradiance
Spherical source of diameter, d_S, not filling receiver field of view	$P_B = \dfrac{\pi \tau_a \tau_r \lambda_i d_S{}^2 d_R{}^2}{16 R^2} \mathcal{W}(\lambda)$	Spectral radiant emittance
	$P_B = \dfrac{\pi^2 \tau_a \tau_r \lambda_i d_S{}^2 d_R{}^2}{16 R^2} \mathcal{N}(\lambda)$	Spectral radiance
	$P_B = \dfrac{\pi^2 \tau_a \tau_r \lambda_i d_S{}^2 d_R{}^2 h f_c}{16 R^2} \mathcal{Q}(\lambda)$	Photon spectral radiance
Extended source filling receiver field of view, θ_R	$P_B = \dfrac{\pi \tau_a \tau_r \lambda_i \theta_R{}^2 d_R{}^2}{4} \mathcal{W}(\lambda)$	Spectral radiant emittance
	$P_B = \dfrac{\pi^2 \tau_a \tau_r \lambda_i \theta_R{}^2 d_R{}^2}{4} \mathcal{N}(\lambda)$	Spectral radiance
	$P_B = \dfrac{\pi^2 \tau_a \tau_r \lambda_i \theta_R{}^2 d_R{}^2 h f_c}{4} \mathcal{Q}(\lambda)$	Photon spectral radiance

Table 6-2. EXPRESSIONS FOR BACKGROUND RADIATION POWER AT DETECTOR SURFACE

radiation. If a point source appears in an extended radiant field, the spectral radiance and irradiance contributions to the total background radiation power must be combined. In all situations the system designer must guard against the incorrect use of the spectral radiance if the source does not fill the receiver field of view, or against the use of the spectral irradiance in the opposite case.

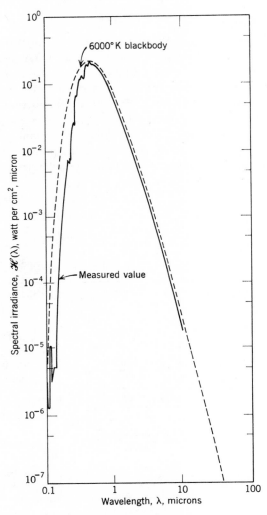

Figure 6-5 *Comparision of sun's spectrum and spectrum of 6000° K blackbody* [6-6]

6.3 BACKGROUND RADIATION SOURCES

The following discussion presents a capsule summary of the characteristics of various background radiation sources.

Sun

Sunlight can be regarded as being radiation from a blackbody source of 6000° K. Figure 6-5 compares the measured spectral density of the sun above the earth's atmosphere and the 6000° K blackbody curve [6-6]. Reception with the sun as a direct background is usually not feasible because of its high brightness.

Sky

The sky presents a background radiance due to scattering of incident radiation and to emission by atmospheric particles as a result of absorption of incident radiation. Figure 6-6 illustrates the measured spectral radiance of the sky under clear sky daytime conditions [6-7]. Radiance for sunlit

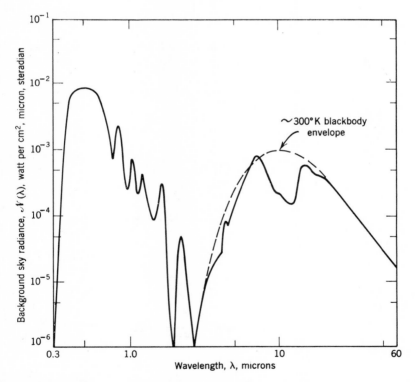

Figure 6-6 *Diffuse component of typical background radiance from sea level, zenith angle 45°, excellent visibility* [6-7]

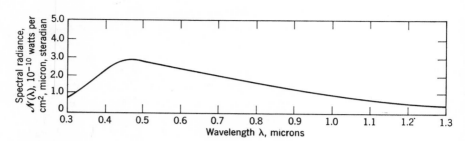

Figure 6-7   *Nighttime sky radiance from zenith due to zodiacal light, galactic light, and scattered starlight* [6-8] (*Courtesy of National Bureau of Standards*)

clouds is approximately an order of magnitude larger. At night, incident sky radiation is due to starlight, zodiacal light, galactic light, air glow, and scattered light from these sources. Figure 6-7 illustrates the magnitude of the nighttime sky radiance [6-8].

Moon and Planets

Background radiation from the moon and planets consists of reflected sunlight and self-emission by the bodies. The reflected radiation spectrum

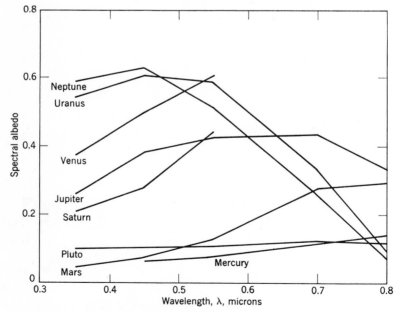

Figure 6-8   *Spectral albedo as a function of wavelength* [6-9]

is the same as the sun's spectrum, a 6000° K blackbody, but the radiation intensity is reduced by the absorption of the body. The ratio of the total reflected radiation to the total incident radiation is called the albedo, *a*, of the body. Figure 6-8 shows the albedo of various planets as a function of wavelength [6-9]. Spectral irradiance curves of the planets are usually plotted by assuming a constant albedo taken at the peak wavelength and, hence, must be modified by the change in albedo with respect to wavelength. Figure 6-9 gives the calculated spectral irradiance of the moon and planets just outside the earth's atmosphere [6-10]. These curves are calculated for the relative positions of the earth to the moon and other planets indicated in Figure 6-10. In general the background radiation from the moon and planets at the surface of the earth is dependent upon the distance of the body from the sun; the distance of the body from the earth; the line-of-sight angle between the sun, body, and earth; and the spectral albedo. Reference [6-11]

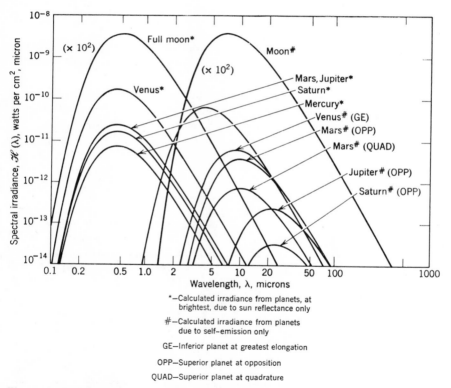

*—Calculated irradiance from planets, at brightest, due to sun reflectance only

#—Calculated irradiance from planets due to self-emission only

GE—Inferior planet at greatest elongation

OPP—Superior planet at opposition

QUAD—Superior planet at quadrature

Figure 6-9   *Calculated planetary and lunar spectral irradiance outside the terrestrial atmosphere [6-10]*

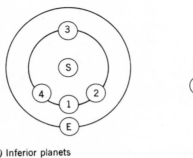
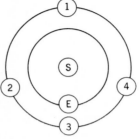

(a) Inferior planets
S = Sun
E = Earth
1 = Position of inferior conjugation
2, 4 = Position of greatest elongation
3 = Position of superior conjugation

(b) Superior planets
S = Sun
E = Earth
1 = Position of conjugation
2, 4 = Position of quadrature
3 = Position of opposition

Figure 6-10   *Arrangements of planets in solar system*

provides parametric data for planetary radiation computations. The blackbody temperature of the planets and the moon and their physical characteristics are listed in Table 6-3 [6-10].

In deep-space optical communication systems, the earth appears as a point source and its irradiance may be assumed constant over the surface area. Figure 6-11 shows the spectral radiant emittance of the earth [6-12]. The irregular shape of the curve is caused by selective wavelength absorption of the atmosphere. For space communication over shorter ranges or for

Planet	Mean Range to Earth ($\times 10^6$ km)	Diameter ($\times 10^3$ km)	Blackbody Temperature (°K)
Mercury	137.8	4.8	613
Venus	103.3	12.4	235
Mars	78.3	6.8	217
Jupiter	628.3	142.8	138
Saturn	1277.0	120.8	123
Uranus	2720.0	47.6	90
Moon	0.384	3.45	373
Earth	—	12.76	300

Table 6-3. BLACKBODY TEMPERATURE OF PLANETS AND MOON

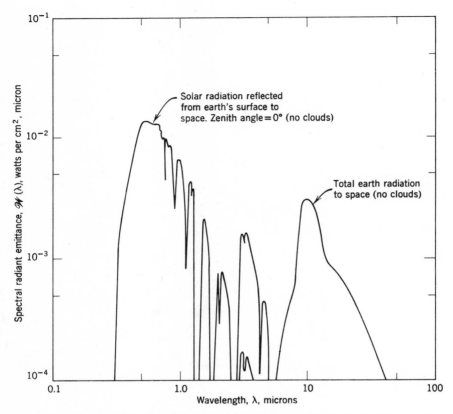

Figure 6-11   *Spectral radiant emittance of the earth* [6-12]

point-to-point communication on the earth, the spectral radiance of various backgrounds such as sea water, plowed fields, vegetation, and rocky terrains, must be considered.

Starlight

The spectral irradiance of the brightest stars just outside the earth's atmosphere is illustrated in Figure 6-12 [6-10]. A visual magnitude scale, M_v, on the figure indicates the brightness of the stars relative to a reference given by the relation

$$M_v = -2.5 \log \left[\frac{\mathcal{H}(\lambda)_{max}}{3.1 \times 10^{-13} \text{ watt/cm}^2} \right] \tag{6-23}$$

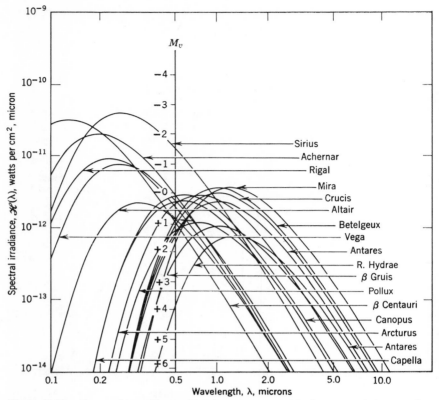

Figure 6-12   *Spectral irradiance of brightest stars outside the terrestrial atmosphere* [*6-10*]

REFERENCES

6-1. Merrit, T. P. and Hall, F. F., Jr. "Black Body Radiation." *Proceedings IRE*, **47** (9), 1435–1442, Sept. 1959.

6-2. Wolfe, W. L. ed. *Handbook of Military Infrared Technology.* Office of Naval Research, U.S. Government Printing Office, Washington, 1965.

6-3. Kruse, P. W. McGlauchlin, L. D., and McQuistan, R. B. *Elements of Infrared Technology: Generation, Transmission, and Detection.* John Wiley, New York, 1962.

6-4. Jamieson, J. A. et al. *Infrared Physics and Engineering.* McGraw-Hill, New York, 1963.

6-5. Smith, R. A., Jones, F. E., and Chasmer, R. P. *The Detection and Measurement of Infra-Red Radiation.* Oxford Press, England, 1957.

6-6. Malitson, H. H. "The Solar Energy Spectrum." *Sky and Telescope,* **29** (4), 162–165, Mar. 1965.

6-7. "Parametric Analysis of Microwave and Laser Systems for Communication and Tracking," Hughes Aircraft Co., Report No. P67-09, Dec. 1966.

6-8. Roach, F. E. "Manual of Photometric Observations of the Airglow During the IGY." National Bureau of Standards Report.

6-9. Kuiper, G. E. and Middlehurst, B. M. ed. *The Solar System, Vol. III, Planets and Satellites.* Univ. of Chicago Press, 1961.

6-10. Ramsey, R. C. "Spectral Irradiance from Stars and Planets Above the Atmosphere From 0.1 to 100 Microns." *Applied Optics,* **1** (4), 465–471, July 1962.

6-11. Meisenholder, G. W. Jet Propulsion Laboratory Report No. 32-361, Nov. 1962.

6-12. Goldberg, I. L. "Radiation From Planet Earth," U.S. Army Signal Research and Development Laboratory Report 2231, AD-266-790, Sept. 1961.

chapter 7

ATMOSPHERIC PROPAGATION

A laser signal propagating through the earth's atmosphere is subject to attenuation due to absorption of radiation by atmospheric constituents and to scattering by particles in the atmosphere. Furthermore, the shape, direction, and electromagnetic properties of a laser beam are affected by atmospheric turbulence.

A complete parametric description of the intertwining effects of the atmosphere on laser signal propagation for all system-operating conditions is beyond the scope of this chapter. The approach taken here is to provide a theoretical basis of the atmospheric propagation characteristics, pertinent examples, and references for further data.

7.1 ATMOSPHERIC ATTENUATION

Atmospheric attenuation may be described by the exponential law of attenuation. The atmospheric transmissivity is

$$\tau_a = \exp\{-\alpha_a L\} \tag{7-1}$$

where L is the atmospheric transmission pathlength and where the attenuation coefficient, α_a, often called the extinction coefficient, is equal to the sum of the absorption and scattering coefficients, α_b and α_s. Thus the atmospheric transmissivity can be factored as the product of the absorption and scattering transmissivities

$$\tau_a = \tau_b \tau_s \tag{7-2}$$

where τ_b is the absorption transmissivity and τ_s is the scattering transmissivity.

Absorption in the atmosphere is due to molecular constituents such as water vapor, carbon dioxide, and ozone. Figure 7-1 illustrates the transmissivity of each of these substances [7-1]. These curves were measured with relatively low spectral resolution. In actuality, the atmospheric absorption varies drastically as a function of wavelength since molecular absorption is basically a line rather than band phenomena. As an illustration, Figure 7-2 shows the measured transmissivity of the atmosphere near the ruby laser

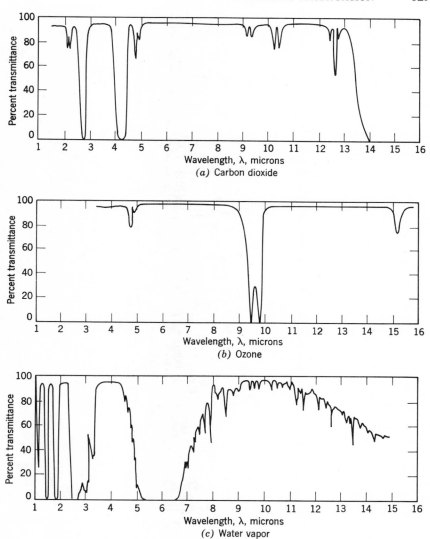

Figure 7-1 *Transmissivities of atmospheric constituents* [7-1]

wavelength [7-2]. This figure points out the desirability of a judicious choice of carrier frequency and also the necessity of stabilizing the laser carrier frequency in order to avoid regions of high absorption. Gross predictions of the atmospheric absorption can be made by estimating the amount of molecular absorbants in the transmission path. Such estimates are difficult

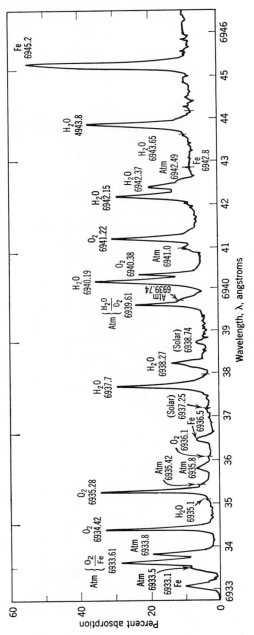

Figure 7-2 Atmospheric absorption near ruby laser wavelength [7-2]

because the molecular concentration of the atmosphere is dependent upon temperature and pressure which vary with geographical location, altitude, and weather. References [7-3] and [7-4] contain tables that give the atmospheric absorption transmissivity as a function of transmission wavelength and amount of water vapor, ozone, or carbon dioxide in the atmospheric path.

Atmospheric scattering, when due to molecular sized particles, is called Rayleigh scattering; when due to particles such as smoke and fog, large comparable to the transmission wavelength, it is called Mie scattering. Rayleigh scattering is predominant at the lower wavelengths, while Mie scattering is much less wavelength dependent.

The attenuation coefficient due to Rayleigh scattering, α_{SR}, is given by the formula

$$\alpha_{SR} = 0.827NA_p{}^3\lambda_c{}^{-4} \tag{7-3}$$

where N is the number of particles per unit volume in the path and A_p is the cross-sectional area of a scattering particle [7-5, 7-6]. In the formula the particle area and wavelength are in units of centimeters. Since the Rayleigh scattering coefficient is inversely proportional to λ^4, short wavelength light is scattered much more than light of a longer wavelength. This accounts for the blue color of the daylight sky. The shorter wavelength components of sunlight are scattered to the ground more than the longer wavelength components. In most situations, Rayleigh scattering is much less than Mie scattering and it may be neglected for practical purposes.

Mie scattering is described by the following empirical relation

$$\alpha_{SM} = \frac{3.91}{\mathscr{V}} \left[\frac{\lambda_c}{0.55}\right]^{-0.585\mathscr{V}^{1/3}} \tag{7-4}$$

where α_{SM} is the Mie scattering coefficient, $\mathscr{V}$ is the visual range in kilometers, and the wavelength and pathlength are expressed in microns and kilometers, respectively [7-1]. Visual range measurements for many localities are reported daily by the United States Weather Bureau. As an example of the use of the Mie scattering formula, consider the transmission of a laser carrier of wavelength, λ_c, over a 1-km path when the visual range is 5 km. The Mie scattering coefficient and scattering transmissivity are listed below for several transmission wavelengths.

λ_c (micron)	α_{SM} (km^{-1})	τ_s
0.5	0.860	0.42
1.0	0.430	0.65
2.0	0.215	0.81
5.0	0.086	0.92
10.0	0.043	0.96

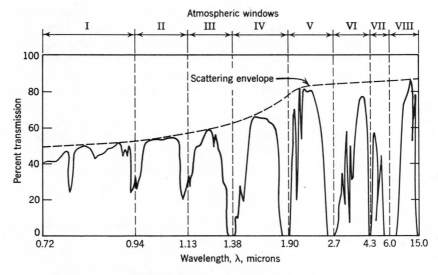

Figure 7-3   *Transmissivity of the atmosphere [7-3]*

Figure 7-3 presents measurements of the atmospheric transmissivity as a function of wavelength, considering both absorption and scattering [7-3]. The decrease in transmissivity below two microns is due primarily to scattering. Selective absorption at the infrared frequencies accounts for the sharp dips in the transmissivity curve. These sharp dips serve to define eight regions of relatively high transmissivity called atmospheric windows.

7.2   ATMOSPHERIC TURBULENCE

Part of the sunlight incident upon the earth's surface is absorbed, causing the surface air layer to be heated by the earth. This sheet of warmed air becomes less dense and rises to mix turbulently with the surrounding cooler air. Air temperature thus varies from point to point in the atmosphere in a random manner. The temperature fluctuations are a function of altitude and wind speed.

The index of refraction of air is dependent upon its temperature. When a light beam traverses a region in which there is a temperature change in the air, the beam is partially or totally deviated, depending upon the relative sizes of the beam and the temperature inhomogeniety. This interaction of the laser beam with the turbulent medium leads to random amplitude and phase variations of the laser carrier. The consequences of atmospheric turbulence on laser communications are listed below [7-7].

Beam steering—angular deviation of the beam from the line-of-sight path, causing the beam to miss the receiver.

Image dancing—variations in the beam-arrival angle, causing the focus point to move in the image plane.

Beam spreading—small angle scattering, increasing the beam divergence and causing a decrease in spatial power density at the receiver.

Beam scintillation—smallscale destructive interference within the beam cross section, causing variations in the spatial power density at the receiver.

Spatial coherence degradation—losses in phase coherence across the beam phase fronts, degrading the photomixing performance.

Polarization fluctuations—variations in the polarization state.

It is convenient to consider a turbulent medium to be composed of discrete blobs, each of which is homogeneous but of different refractive index than its neighbors. An inhomogeneity dimension l, is associated with each blob. The smallest and largest inhomogeneities are characterized by the dimensions l_o and L_o, respectively.

The effect of atmospheric turbulence depends upon the relative sizes of the beam diameter, d_B, and the inhomogeneity dimension, l. If $d_B/l \ll 1$, the major effect of turbulence is to deflect the beam as a whole. At long ranges the beam appears to execute a two-dimensional random walk in the receiver plane. For $d_B/l \approx 1$, the inhomogeneities act as lenses which focus or defocus all or parts of the beam, imparting a granular structure to the beam cross section. If $d_B/l \gg 1$, small portions of the beam are independently diffracted and the beam phase fronts are badly distorted. Figure 7-4 contains sketches that illustrate turbulence effects for the extreme cases when $d_B/l \gg 1$ and $d_B/l \ll 1$.

For communications from a deep-space transmitter to an earth receiver, $d_B/l \gg 1$. The principal turbulence effects are then beam spreading, beam scintillation, and spatial coherence degradation. In the opposite case for an earth transmitter and deep-space receiver, $d_B/l \ll 1$. Under this condition, image dancing and beam steering are the predominant effects of atmospheric turbulence.

The degree of atmospheric turbulence and its relationship to the optical properties of the atmosphere may be characterized by Tatarski's structure constant for refractive index fluctuations, $C_n(r)$ [7-8]. The value of the structure constant varies with altitude and time of day. Typical values for daytime conditions near the earth surface are listed below [7-7].

Weak turbulence—$C_n = 8 \times 10^{-9} \text{ m}^{-1/3}$.
Intermediate turbulence—$C_n = 4 \times 10^{-8} \text{ m}^{-1/3}$.
Strong turbulence—$C_n = 5 \times 10^{-7} \text{ m}^{-1/3}$.

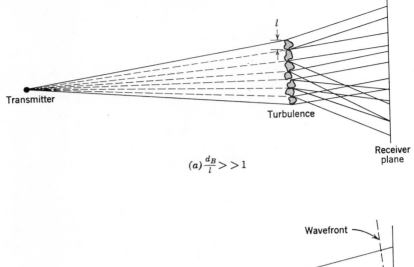

$$(a)\frac{d_B}{l}>>1$$

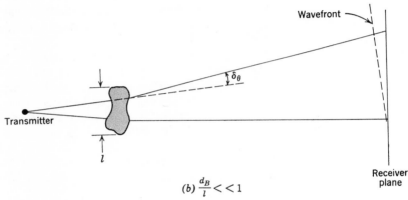

$$(b)\frac{d_B}{l}<<1$$

Figure 7-4 *Atmospheric turbulence effects as a function of beam diameter and turbulence dimension*

In Tatarski's model for the turbulent atmosphere, the electric field amplitude variations of a beam are given by a log normal distribution. The value of the mean square amplitude fluctuations is

$$\overline{\left[\ln\frac{A}{A_i}\right]^2} = 0.56\left(\frac{2\pi}{\lambda_c}\right)^{7/6}\int_0^L C_n^2(r)z^{5/6}\,dz \tag{7-5}$$

where A is the instantaneous electric field amplitude and A_i is the initial, or reference, electric field amplitude [7-8]. Propagation is over a path length, L, in the turbulent medium satisfying the relationship $(\lambda_c L)^{1/2} \gg l_o$. If the structure constant does not change appreciably along the beam path, the fluctuation in the logarithmic level of the beam amplitude reduces to

$$\overline{\left[\ln\left(\frac{A}{A_i}\right)\right]^2} = 0.31 C_n{}^2 \left(\frac{2\pi}{\lambda_c}\right)^{7/6} (L)^{11/6} \qquad (7\text{-}6)$$

Figure 7-5 is a plot of the rms logarithmic amplitude fluctuation as a function of propagation pathlength and wavelength [7-7].

The fluctuation in phase between points in the beam separated by the distance, ρ, as postulated by Tatarski, follows a Gaussian distribution with a variance

$$\sigma_\Phi{}^2(\rho) = \begin{cases} 1.46\left(\frac{2\pi}{\lambda_c}\right)^2 \rho^{5/3} \int_0^L C_n{}^2(r)\,dz & \text{for } l_o < \rho < (\lambda_c L)^{1/2} \quad (7\text{-}7) \\[2ex] 2.91\left(\frac{2\pi}{\lambda_c}\right)^2 \rho^{5/3} \int_0^L C_n{}^2(r)\,dz & \text{for } L_o > \rho > (\lambda_c L)^{1/2} \quad (7\text{-}8) \end{cases}$$

Physically, the phase fluctuations are caused by either wavefront distortion over a lateral beam distance ρ such that $\sigma_\Phi(\rho) > \pi$ or a plane wavefront being tilted through an angle

$$\sigma_\theta = \frac{\sigma_\Phi(\rho)}{(2\pi/\lambda_c)\rho}$$

The first cause leads to beam spreading, beam scintillation, and spatial coherence degradation, while the second contributes to beam steering and image dancing.

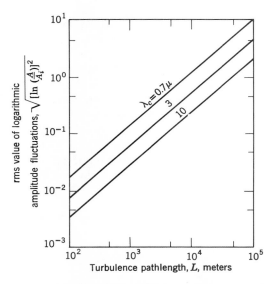

Figure 7-5 *The rms value of logarithmic amplitude fluctuations due to intermediate atmospheric turbulence [7-7]*

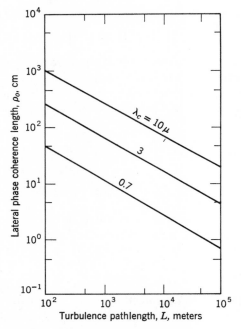

Figure 7-6   Lateral phase coherence length for intermediate turbulence [7-7]

It is useful to define a lateral phase coherence length, ρ_0, by the condition that $\sigma_\Phi(\rho_0) = \pi$. The spatial coherence of the beam is reduced appreciably over lateral beam distances greater than ρ_0. Figure 7-6 is a plot of the lateral phase coherence length for intermediate turbulence [7-7].

Tatarski's turbulent atmosphere model is based upon a perturbation method, called Rytov's method, for the solution of the electromagnetic wave equation in the turbulent atmosphere [7-8, 7-9]. Recent theoretical and experimental studies indicate some questions as to the validity of Rytov's method for certain turbulence conditions [7-10]. With this qualification in mind, the following sections present a summary of the analyses of atmospheric turbulence effects predictated upon the Tatarski model.

Beam Steering

The rms angular deviation of a single ray of a bundle of rays comprising a beam is [7-7]

$$\delta_\theta = \frac{\sigma_\theta}{\sqrt{2}} \tag{7-9}$$

If the beam dimension is much less than the inhomogeneity dimension (i.e., $d_B/l \ll 1$), the entire beam will be deviated by the angle δ_θ. Figure 7-7 is a plot of the rms beam deviation angle, δ_θ, as a function of the propagation

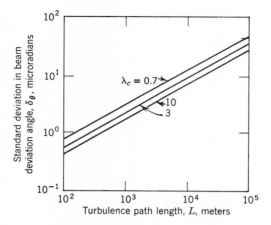

Figure 7-7 *Standard deviation in beam deviation angle of a phase coherent portion of a laser beam due to intermediate atmospheric turbulence* [7-7]

wavelength and range [7-7]. Compensation for beam steering may be effected simply by increasing the beam divergence angle so that even if the beam is deviated by turbulence, the receiver will still be illuminated. The penalty paid, of course, is a reduction in spatial power density in the receiver plane.

Image Dancing

The rms displacement of the focus point from the focal point of the receiver antenna is equal to $F\sigma_\theta$ where F is the focal length of the receiver optical antenna. Figure 7-8 is a plot of the standard deviation in beam arrival

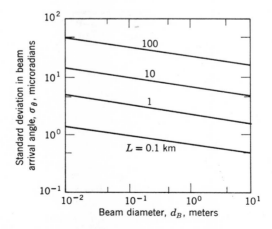

Figure 7-8 *Standard deviation in beam arrival angle due to intermediate atmospheric turbulence* [7-7]

angle as a function of the beam diameter, d_B, and propagation pathlength [7-7]. In most cases the deviation in beam arrival angle is small and image dancing is negligible. The method of compensation, if image dancing is of concern, is to provide a larger detector area which increases the receiver field of view. Increasing the field of view, however, results in a greater amount of background radiation entering the receiver.

Beam Spreading

Beam spreading occurs when the beam wavefront is distorted such that $\sigma_\Phi(\rho) > \pi$ over some lateral beam distance, ρ. Corresponding to each phase coherent portion of the beam, defined by a phase coherence length, ρ_0, there is a ray deviation angle δ_θ which is relatively independent from one coherence area of the beam to another. Hence the beam energy is randomly spread in the receiver plane.

An estimate of the reduction in spatial power density at the receiver due to beam spreading has been derived by Davis [7-7]. The fraction of the total power incident upon a receiver antenna after beam spreading is shown in Figure 7-9 as a function of the sizes of the transmitter and receiver antennas.

Beam Scintillation

Figure 7-10 illustrates the far-field cross section of a laser beam which has passed through a turbulent atmosphere [7-11]. The destructive and constructive regions of self-interference of the beam are characterized by the light and dark patches in the photograph. Analysis of this particular photograph has shown that 70% of the area of the beam has an intensity less than

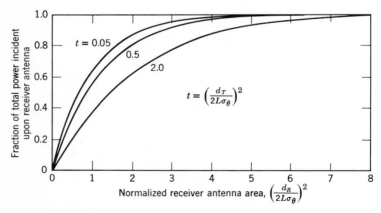

Figure 7-9 *Effect of beam spreading due to intermediate atmospheric turbulence* [7-7]

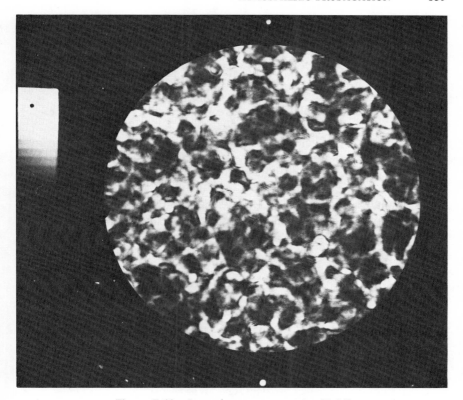

Figure 7-10   *Laser beam cross section* [7-11]

the average intensity. The highest peaks in intensity are about four times the average.

Beam scintillation creates a random fading in the carrier beam intensity which can lead to a decrease in the signal-to-noise ratio of an optical receiver. There is a possibility of spectral interference between the carrier information modulation and the beam intensity fluctuations.

For a direct detection receiver the average photodetector signal current can be written in terms of the logarithmic amplitude fluctuation ratio given by Equation 7-5. Fried has derived an expression for the carrier signal variance in terms of the spatial covariance of the logarithmic amplitude ratio [7-12]. Results indicate that the signal variance is approximately inversely proportional to the receiver antenna area. If the receiver antenna diameter is a small fraction of the laser beam diameter in the receiver plane, statistical variations in the photodetector signal current may be quite large in the presence of turbulence. As the antenna diameter-to-laser beam diameter ratio increases, the magnitude of the variations decreases. This is called the aperture averaging effect. In the limit as the receiver encompasses the entire beam in the receiver plane, there is no variation in the signal current. Thus, in a direct detection receiver, increasing the antenna diameter not only increases the average signal level but also decreases the ripple in the average signal level.

In a heterodyne detection receiver the photodetector signal current at the intermediate frequency filter output can be expressed in terms of the logarithmic amplitude fluctuation ratio and the fluctuating carrier phase. The variance in the signal current has been found to depend upon an aperture averaging factor and a phase degradation factor [7-13]. As the aperture diameter increases, the effect of intensity fluctuations decreases, but the contribution to the variance from phase fluctuations increases. Thus, there is some compromise value of receiver antenna diameter which minimizes the signal variance. In most situations the best choice for the antenna diameter is a value approximately equal to the phase coherence dimension, r_0, defined in Equation 7-11.

Spatial Coherence Degradation

Atmospheric turbulence distorts the phase fronts of a plane wave propagating through the atmosphere such that over a receiving plane the amplitude and phase of the wave vary from point to point [7-14]. This results in a decrease in the photodetector output signal of a heterodyne or homodyne detection optical receiver. Since the receiver noise is independent of the signal level under proper operating conditions, the signal-to-noise ratio and, hence, system performance are degraded [7-15, 7-16].

Fried has derived an expression for the degradation in the signal-to-noise ratio for an optical heterodyne receiver in terms of the relative sizes of the receiver antenna, d_R, and a phase coherence dimension, r_0 [7-17]. The signal-to-noise ratio degradation, $\psi(d_R/r_0)$, is

$$\psi\left(\frac{d_R}{r_0}\right) = \frac{32}{\pi} \left(\frac{d_R}{r_0}\right)^2 \int_0^1 \frac{u}{2} \left[\cos^{-1}(u) - u\sqrt{1 - u^2}\right] \exp\left\{-3.44\left(\frac{d_R}{r_0}\right)^{5/3} u^{5/3}\right\} du$$

$$(7\text{-}10)$$

Figure 7-11 is a plot of $\psi(d_R/r_0)$ which indicates that very little improvement will result from making the receiver antenna diameter much larger than r_0 [7-17].

The quantity r_0 is related to the transmission wavelength, zenith angle, θ, and receiver altitude, H_0, in meters by the relation

$$r_0 = 0.05(\lambda_c)^{6/5} \cos^{3/5}\theta \left[\frac{\Gamma(2/3)}{\Gamma\left(2/3 \frac{H_0}{3200}\right)}\right]^{3/5}$$

$$(7\text{-}11)$$

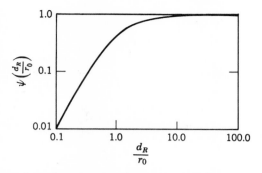

Figure 7-11   *Heterodyne detection signal-to-noise ratio degradation as a function of receiver antenna diameter [7-13]*

where $\Gamma(x, y)$ is the incomplete Gamma function [7-17]. This expression is valid for daytime operation. At night, r_0 is about twice as large as given by Equation 7-11. Figure 7-12 is a plot of r_0 as a function of receiver altitude, zenith angle, and transmission wavelength [7-17]. The effects of receiver altitude and zenith angle on r_0 are relatively weak. However, there is a large increase in r_0 as a function of wavelength due to the $(\lambda_c)^{6/5}$ dependence. For example, r_0 is thirty-seven times as large at 10 μ as at 0.5 μ. The coherence dimension can also be related to the atmospheric structure constant, C_n, and propagation pathlength, L. For a horizontal propogation path.

$$r_0 = 1.2 \times 10^{-8}(\lambda_c)^{6/5}(L)^{-3/5}(C_n)^{-6/5} \tag{7-12}$$

The dependence of r_0 on wavelength, range, and the value of the structure constant is shown in Figure 7-13 [7-17].

Polarization Fluctuations

The electric field of a wave entering a region of turbulence can be broken up into components E_{X1} and E_{Y1} in the plane of incidence [7-18, 7-19]. The polarization angle of the incident wave is

$$\Xi = \tan^{-1}\frac{E_{Y1}}{E_{X1}} \tag{7-13}$$

After leaving the turbulence region, the electric field components are changed to E_{X2} and E_{Y2} due to a change in the index of refraction of the medium. The polarization angle is then

$$\Xi + \Delta\Xi = \tan^{-1}\frac{E_{Y2}}{E_{X2}} \tag{7-14}$$

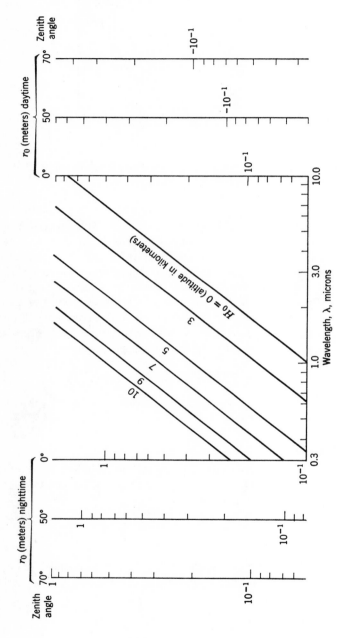

Figure 7-12 Dependence of r_0 on receiver altitude, zenith angle, and transmission wavelength [7-13]

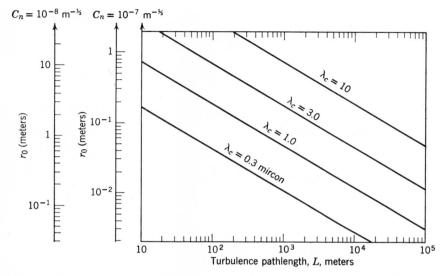

Figure 7-13 *Dependence of r_0 on transmission wavelength, turbulence pathlength, and turbulence structure constant [7-15]*

Saleh has developed an expression for the mean square change in the polarization angle [7-20]. For an isotropic atmosphere the rms polarization change is

$$\sqrt{\overline{[\Delta \Xi]^2}} = 2\pi \overline{[\Delta n]^2} \frac{L}{L_o} \tag{7-15}$$

where $\overline{[\Delta n]^2}$ is the mean square change in index of refraction due to thermal variations. The index of refraction change has been empirically determined to be related to the altitude of observation, H_0, in meters by the relation [7-18]

$$\overline{[\Delta n]^2} = 10^{-12} \exp\left\{-\frac{H_0}{1600}\right\} \tag{7-16}$$

As an example of transmission over a low-altitude ($H_o \approx 0$), horizontal path of 10^4 m with a turbulence dimension of $L_o = 1$ m, the rms change in polarization angle, $\sqrt{\overline{[\Delta \Xi]^2}}$, in on the order of 10^{-8} rad. Experiments verify that polarization fluctuations do not appear to be a problem for laser propagation [7-20].

REFERENCES

7-1. Kruse, P. W., McGlauchlin, L. D., and McQuistan, R. B. *Elements of Infrared Technology: Generation, Transmission, and Detection.* John Wiley, New York, 1962.

7-2. Long, R. K. "Atmospheric Attenuation of Ruby Lasers." *Proceedings IEEE,* **51** (5), 859–860, May 1963.

7-3. Langer, R. M. "Effects of Atmospheric Water Vapour on near Infrared Transmission at Sea Level. Report on Signal Corps Contract DA-36-039-SC-723351. J. R. M. Bege Co., Arlington, Mass., May, 1957.

7-4. Jamieson, J. A. et al. *Infrared Physics and Engineering.* McGraw-Hill, New York, 1963.

7-5. Van De Hulst, H. C. *Light Scattering by Small Particles.* John Wiley, New York, 1957.

7-6. Wolfe, W. L., Ed. *Handbook of Military Infrared Technology.* Office of Naval Research, U.S. Government Printing Office, Washington, 1965.

7-7. Davis, J. I. "Consideration of Atmosphere Turbulence in Laser Systems Design." *Applied Optics,* **5** (1), 139–147, Jan. 1966.

7-8. Tatarski, V. I. *Wave Propagation in Turbulent Medium.* McGraw-Hill, New York, 1961.

7-9. Chernov, L. A. *Wave Propagation in a Random Medium.* McGraw-Hill, New York, 1960.

7-10. Strohbehn, J. W. "Line-of-Sight Wave Propagation Through the Turbulent Atmosphere." *Proceedings IEEE,* **56** (8), 1301–1318, Aug. 1968.

7-11. Deitz, P. H. "Near Earth Propagation of Optical Beams." *Electronic Communicator,* **2** (5), 15, Sept./Oct. 1967.

7-12. Fried, D. L. "Aperture Averaging of Scintillation." *Journal of Optical Society of America,* **57** (1), 169–175, Jan. 1967.

7-13. Fried, D. L. "Atmospheric Modulation Noise in an Optical Heterodyne Receiver." *IEEE Journal of Quantum Electronics,* **QE-3** (6), 213–221, June 1967.

7-14. Fried, D. L. "Statistics of a Geometric Representation of Wavefront Distortion." *Journal Optical Society of America,* **55** (11), 1427–1435, Nov. 1965.

7-15. Gardner, S. "Some Effects of Atmospheric Turbulence on Optical Heterodyne Communications." *1964 IEEE Convention Record,* Part 6, 337–342, 1964.

7-16. Goldstein, I., Miles, P. A., and Chabot, A. "Heterodyne Measurements of Light Propagation Through Atmospheric Turbulence." *Proceedings IEEE,* **53** (9), 1172–1180, Sept. 1965.

7-17. Fried, D. L. "Optical Heterodyne Detection of an Atmospherically Distorted Signal Wavefront." *Proceedings IEEE,* **55** (1), 57–67, Jan. 1967.

7-18. Hodara, H. "Laser Wave Propagation Through the Atmosphere." *Proceedings IEEE,* **54** (3), 368–375, Mar. 1966.

7-19. Saleh, A. A. M. and Hodara, H. "Comments on Laser Wave Propagation Through the Atmosphere." *Proceedings IEEE Letters,* **55** (7),1209, July 1967.

7-20. Saleh, A. A. M. "An Investigation of Laser Wave Depolarization due to Atmospheric Transmission." *IEEE Journal of Quantum Electronics,* **QE-3** (11), 540–543, Nov. 1967.

chapter 8

DETECTION NOISE

Noise in the detection process of a laser communication system arises from radiation entering the receiver and from internally generated noise. The major types of detection noise are listed below.

Internal Noise	External Noise
Thermal noise	Photon fluctuation shot and
Flicker noise	generation-recombination noise
Current noise	Radiation intensity fluctuation noise
Dark current shot noise	Phase noise

8.1 THERMAL NOISE

Thermal or Johnson noise [8-1, 8-2] is caused by thermal fluctuations of electrons in a resistor. Consider the "noisy" resistor, R_L, of Figure 8-1 which is connected in parallel with a capacitor, C. In a practical detector, R_L may represent the load resistor and C the shunt capacitance of the detector. The situation in which the detector contains resistive elements will be considered later.

The average energy stored in the capacitor due to the noisy resistor alone ($\frac{1}{2}C\widetilde{v^2}$) may be equated to the thermodynamic energy of the system ($\frac{1}{2}kT$) by the principle of the equipartition of energy for a system in thermal equilibrium.† Thus,

$$\tfrac{1}{2}C\widetilde{v^2} = \tfrac{1}{2}kT \tag{8-1}$$

or

$$\widetilde{v^2} = \frac{kT}{C} \tag{8-2}$$

---

† This derivation assumes the thermodynamic energy to be $\frac{1}{2}kT$. In general the system energy is $\frac{1}{2}hf/(\exp hf/kT - 1)$ which reduces to $\frac{1}{2}kT$ for the frequencies of interest.

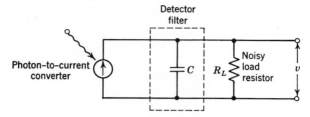

Figure 8-1   *Photodetector with capacitor filter and noisy load resistor*

where:

$$k = \text{Boltzmann's constant}$$

$$T = \text{resistor temperature}$$

$$\widetilde{v^2} = \text{mean square resistor voltage}$$

The thermal noise power is then

$$N_T = \frac{\widetilde{v^2}}{R_L} = \frac{kT}{R_L C} \tag{8-3}$$

If the detector bandwidth is defined as the reciprocal of the time constant ($B \equiv 1/R_L C$), the thermal noise power expression for a resistor-capacitor filter can be written as

$$N_T = kTB \tag{8-4}$$

The thermal noise associated with the circuit of Figure 8-1 may also be described by a thermal noise generator producing a noise voltage, as shown in Figure 8-2*a*, or by the equivalent current generator in Figure 8-2*b*. The electron fluctuations in the noisy resistor occur in very short times compared to the response time of the detector. Thus, the spectrum of the thermal noise source may be assumed to be flat.† Since the output noise spectral density of a linear system is the product of the input noise spectral density and the square of the absolute value of the Fourier transform of the impulse response of the system, the thermal noise voltage fluctuations may be written as

$$\widetilde{v^2} = G_{v_T} R_L \int_{-\infty}^{\infty} |H(f)|^2 \, df \tag{8-5}$$

where G_{v_T} is the thermal noise power spectral density in voltage units.‡

---

† This assumption, of "White noise," would lead to an infinite noise power over all frequencies. However, at high frequencies the energy per degree of freedom decreases and the power associated with the thermal noise fluctuations is bounded.

‡ If the noise power spectral density is given in voltage units, G_v, the noise power across a resistor, R_L, per unit bandwidth is G_v/R_L. Likewise, if the noise power spectral density is given in current units, G_i, the noise power per unit bandwidth is $G_i R_L$.

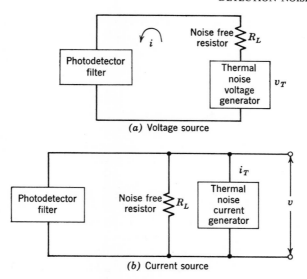

Figure 8-2   *Thermal noise models of photodetector*

The Fourier transform of the impulse response of the $R_L C$ network is

$$H(f) = \frac{1}{1 + j2\pi R_L C f} \tag{8-6}$$

Equating the right side of Equation 8-5 to Equation 8-2 and performing the integration yield the two-sided thermal noise power spectral density

$$G_{v_T} = 2kTR_L \tag{8-7}$$

Equation 8-7 gives the power spectrum of a noisy resistor which can be connected to any detector in any network, as evidenced by the fact that the expression is independent of C and depends only on the value of the noisy resistor and its temperature.

The mean square voltage fluctuation at frequencies in pass bands between $-f_2$ to $-f_1$ and f_1 to f_2 is

$$\overline{v_T^2 (\Delta f)} = 4kTR_L \, \Delta f \tag{8-8}$$

where $\Delta f \equiv f_2 - f_1$. This leads to the treatment of thermal noise as being characterized by an open circuit rms voltage of

$$[\overline{v_T^2 (\Delta f)}]^{1/2} = (4kTR_L \, \Delta f)^{1/2} \tag{8-9}$$

or its rms current equivalent

$$[\overline{i_T^2 (\Delta f)}]^{1/2} = \left(\frac{4kT \, \Delta f}{R_L}\right)^{1/2} \tag{8-10}$$

Care must be taken in the application of these equations to electrical networks. It must be remembered that Δf is not necessarily the bandwidth of the circuit but merely a positive frequency interval over which Equations 8-9 and 8-10 apply.

As an example of the application of these equations, consider a thermal noise current generator (Figure 8-2b) connected to a filter consisting of a shunt capacitor. The total thermal noise power is the integral over all positive frequencies of the product of the mean square, thermal noise current spectral density and the real part of the impedance of the parallel combination of R_L and C. Thus, the thermal noise power is

$$N_T = \int_0^\infty \left[\frac{4kT}{R_L}\right]\left[\frac{R_L}{(2\pi R_L Cf)^2 + 1}\right] df = \frac{kT}{R_L C} \tag{8-11}$$

as derived in Equations 8-3 and 8-4 where the circuit bandwidth was defined to be the inverse of the resistor-capacitor constant.

The thermal noise voltage and current sources, v_T and i_T of Figure 8-2, are Gaussian random variables under the conditions of thermal equilibrium [8-3]. Hence the probability distributions of the thermal noise generators are the Gaussian distributions

$$p(i_T) = \frac{1}{\sqrt{2\pi\sigma_{i_T}^2}} \exp\left\{-\frac{i_T^2}{2\sigma_{i_T}^2}\right\} \tag{8-12}$$

and

$$p(v_T) = \frac{1}{\sqrt{2\pi\sigma_{v_T}^2}} \exp\left\{-\frac{v_T^2}{2\sigma_{v_T}^2}\right\} \tag{8-13}$$

where $\sigma_{i_T}^2 = N_T/R_L$ is the variance of the Gaussian thermal noise current source and $\sigma_{v_T}^2 = N_T R_L$ is the variance of the Gaussian thermal noise voltage source.

8.2 DARK CURRENT SHOT NOISE

In a photoemissive or photovoltaic detector, the dark current flowing in the absence of any external photoexcitation causes shot noise. Experimental evidence indicates that dark current electronic emissions are time independent and obey Poisson statistics [8-4]. The probability that the number of dark current electrons emitted in the time period τ is exactly an integer k is given by the Poisson distribution

$$P(U_{D,\tau} = k) = \frac{(\mu_{D,\tau})^k \exp\{-\mu_{D,\tau}\}}{k!} \tag{8-14}$$

where $\mu_{D,\tau} \equiv \tau I_D/q$ is the average number of dark current electrons released by the detector in a time period τ. These k electrons, emitted at random times within the period τ, each carry a unit electronic charge, q, and produce a total current of

$$i_D(t) = \sum_{n=1}^{k} Gq\delta(t - t_n) \qquad \text{for} \quad -\frac{\tau}{2} \leq t \leq \frac{\tau}{2} \tag{8-15}$$

where $\delta(t - t_n)$ is the unit impulse function occurring at time t_n and G is the post detector current gain. The instantaneous detector dark current is a Poisson process with an autocorrelation function given by

$$R_{I_D}(\gamma) = \int_{-\tau/2}^{\tau/2} \cdots \int_{-\tau/2}^{\tau/2} \sum_{k=0}^{\infty} \left[\sum_{n=1}^{k} Gq\delta(t - t_n) \right]$$

$$\times \left[\sum_{m=1}^{k} Gq\delta(t + \gamma - t_m) \right] \frac{1}{\tau^k} P(U_{D,\tau} = k)\, dt_1 \cdots dt_k \tag{8-16}$$

By the methods described in Appendix C, the autocorrelation function may be written in terms of the average number of dark current photoelectrons

$$R_{I_D}(\gamma) = \frac{G^2 q^2}{\tau} \mu_{D,\tau} \delta(\gamma) + \frac{G^2 q^2}{\tau^2} \mu_{D,\tau}^2 \tag{8-17}$$

or in terms of the average dark current

$$R_{I_D}(\gamma) = G^2 q I_D \delta(\gamma) + G^2 I_D^2 \tag{8-18}$$

The noise power spectral density, which is the Fourier transform of the autocorrelation function, is

$$G_{I_D}(f) = G^2 q I_D + G^2 I_D^2 \delta(f) \tag{8-19}$$

The spectral density is thus composed of a flat spectrum ($G_{i_D} = G^2 q I_D$) plus a direct current component. The total noise power at a load resistor, R_L, due to the fluctuations about the mean through a zonal filter of bandwidth B Hz is†

$$N_{H_D} = 2G^2 q I_D B R_L \tag{8-20}$$

This expression is called the Schottky shot noise formula.

Thus far only the dark current emission fluctuations at the cathode surface have been considered. As the electrons move from the cathode surface to the anode, the noise spectrum is modified as a result of the transit time spread of emitted electrons. The transit time-spread effect may be

---

† If the detector filter is not a zonal low pass filter but rather a simple resistor-capacitor filter of bandwidth B, as in Figure 8-1, the shot noise power is

$$N_{H_D} = \frac{G^2 q I_D B R_L}{2}$$

considered equivalent to a low pass filtering operation where the low pass filter has a single time constant equal to the time spread τ_e. The transfer function of the equivalent low pass filter is then

$$H_{\tau_e}(f) = \frac{1}{1 + 2\pi f \tau_e} \tag{8-21}$$

Multiplying the absolute value squared of $H_{\tau_e}(f)$ by the power spectral density at the photodetector surface gives the photodetector output spectral density. For moderate modulation frequencies, τ_e is much less than the reciprocal detector filter bandwidth, and the electron transit time effect is negligible.

8.3 PHOTON FLUCTUATION NOISE

All types of photodetectors are subject to noise caused by emission fluctuations of the optical radiation incident upon a photodetector, even when the mean radiation intensity is constant. This phenomenon, often called photon fluctuation noise, manifests itself as shot noise for a photoemissive detector or as generation-recombination noise for a solid-state detector [8-5]. Let

$$i_R(t) \equiv \sum_{n=1}^{k} Gq\delta(t - t_n) \tag{8-22}$$

represent the photodetector current caused by the random emission of k photoelectrons at times t_n in a total time interval τ due to a general optical radiative source. The autocorrelation function of this process is

$$R_{I_R}(\gamma) = \int_{-\tau/2}^{\tau/2} \cdots \int_{-\tau/2}^{\tau/2} \sum_{k=0}^{\infty} \left[\sum_{n=1}^{k} Gq\delta(t - t_n) \right]$$
$$\times \left[\sum_{n=1}^{k} Gq\delta(t + \gamma - t_m) \right] \frac{1}{\tau^k} P(U_{R,\tau} = k) \, dt_1 \cdots dt_k \tag{8-23}$$

where $P(U_{R,\tau} = k)$ is the probability distribution of photoelectrons emitted by the detector in the time interval τ due to optical radiation incident upon the photodetector. The autocorrelation function reduces to (Appendix C)

$$R_{I_R}(\gamma) = \left[\frac{\delta(\gamma)}{\tau} - \frac{1}{\tau^2} \right] E(U_{R,\tau}) + \frac{1}{\tau^2} E(U_{R,\tau}^2) \tag{8-24}$$

and the spectral density is

$$G_{I_R}(f) = \frac{E(U_{R,\tau})}{\tau} + \left[\frac{E(U_{R,\tau}^2) - E(U_{R,\tau})}{\tau^2} \right] \delta(f) \tag{8-25}$$

where $E(U_{R,\tau})$ and $E(U_{R,\tau}{}^2)$ are the first and second moments of the probability distribution $P(U_{R,\tau} = k)$.

From Chapter 1, the mean and variance of background radiation photo-electron counts are both equal to $\mu_{B,\tau}$, and the spectral density of the background radiation photon fluctuations is

$$G_{I_B}(f) = G^2 q I_B + G^2 I_B{}^2 \delta(f) \tag{8-26}$$

where $I_B = q\mu_{B,\tau}/\tau$ is the average detector current due to background radiation. The probability distribution of photoelectron counts due to an ideal laser is Poisson, as noted by Equation 1-25. The mean and variance of this distribution are both equal to $\mu_{S,\tau}$, and the spectral density of the laser radiation photon fluctuations is then

$$G_{I_S}(f) = G^2 q I_S + G^2 I_S{}^2 \delta(f) \tag{8-27}$$

where $I_S = q\mu_{S,\tau}/\tau$ is the average detector current due to the laser radiation.

Equations 8-26 and 8-27 describe the power spectral densities of the detector current due to photon fluctuations referred to the detector surface. To determine the noise produced by photon fluctuations, it is necessary to determine the power spectral densities of the photon fluctuations referred to the output of the detector.

In a photoemissive detector, each arriving photon liberates a photoelectron with probability η, where η is the detector quantum efficiency. There is a one-to-one correspondence between each arriving photon and each emitted electron. In a photoemissive detector, the power spectral densities of electron emissions about the mean due to laser and background radiation are

$$G_{i_S}(f) = G^2 q I_S \tag{8-28}$$

and

$$G_{i_B}(f) = G^2 q I_B \tag{8-29}$$

These noise power spectral densities are of the same form as the power spectral density of shot noise caused by detector dark current; hence, they are often called the shot noise due to laser and background radiation, respectively.

In a photoconductive detector an arriving photon creates an average of η hole-electron pairs. The simultaneous generation and recombination processes caused by an arriving photon result in noise power spectral densities which are twice as large as those given by Equations 8-26 and 8-27. The resulting noise power spectral densities about the mean for a photoconductive detector are then

$$G_{i_S}(f) = 2G^2 q I_S \tag{8-30}$$

and

$$G_{i_B}(f) = 2G^2 q I_B \tag{8-31}$$

These noise spectra describe a noise source which is called generation-recombination noise by many authors and simply shot noise by others. Lattice vibrations in the photoconductive material cause a modification of the basic generation-recombination noise spectrum. The effect of lattice vibrations can be found by multiplying the generation-recombination noise spectral density by the square of the absolute value of a transfer function describing the lattice variations. This transfer function is dependent upon the fractional ionization of the material and whether it is intrinsic or extrinsic. Jamieson et al. [8-5] give the generation-recombination noise power spectral densities for several lattice arrangements. In most detectors the lattice time constants are short with respect to the reciprocal detector filter bandwidth, and the lattice vibration effects can be ignored.

Each photon arriving at the surface of a photovoltaic detector produces an average of η hole-electron pairs. However, the recombination lifetimes in a photovoltaic detector are so short that the recombination process does not produce significant fluctuations. Thus the noise power spectral densities for a photovoltaic detector are

$$G_{i_S}(f) = G^2 q I_S \qquad (8\text{-}32)$$

$$G_{i_B}(f) = G^2 q I_B \qquad (8\text{-}33)$$

8.4 OTHER NOISE SOURCES

Flicker Noise

Fluctuation in the emission from spots on a vacuum tube cathode creates what is called flicker noise [8-5]. The spectrum of this noise is inversely proportional to frequency for frequencies down to 1 Hz and to the square of the photocathode current, I_K. The power spectral density of flicker noise is

$$G_{i_F}(f) = \alpha_F \frac{G^2 I_K^2}{f} \qquad (8\text{-}34)$$

where α_F is a proportionality constant. The expression cannot hold down to zero frequency since the noise power must be bounded. Flicker noise is associated chiefly with thermionic emission. It is not usually a significant noise factor in photoemissive tubes.

Current Noise

Semiconductor detectors carrying a steady current exhibit a noise effect, called current of $1/f$ noise, which has a one-sided power spectrum inversely proportional to frequency to below 1 Hz [8-5]. The noise spectrum depends upon the square of the post detector amplified current and the operational environment—most notably upon the humidity but not upon the temperature of the device. The physical mechanism producing current noise is believed to be the trapping of charge carriers near the surface of a material. The power spectral density of current noise is

$$G_{i_c}(f) = \alpha_c \frac{G^2 I_P{}^2}{f} \tag{8-35}$$

where α_c is a proportionality constant.

Radiation Intensity Fluctuation Noise

In addition to shot and generation-recombination noise caused by photon time-of-arrival fluctuations, further receiver noise is caused by random variations in the mean intensity of radiation incident upon the detector. Variations in intensity result from intensity fluctuations of the laser carrier, natural pulsations of the solar source or stars, and random intensity modulation of radiation passing through the atmosphere.

Lasers suitable for communications are generally well amplitude stabilized and are not a serious source of radiation intensity fluctuation noise. Background radiation sources usually do not exhibit rapid changes in radiance. However, the atmosphere can be a significant cause of radiation intensity fluctuations.

Atmospheric intensity fluctuation of a laser beam appears as a scintillation or quivering of the intensity of the beam in the receiver plane. Beam scintillation effects have been discussed in Chapter 7. The expressions for the variance in the signal current of a direct or heterodyne detection optical receiver described in Chapter 7 are essentially measures of the radiation intensity fluctuation noise power.

Phase Noise

In a heterodyne or homodyne optical receiver, in which optical mixing occurs, the spectral, or line, shape of the transmitter and local oscillator laser lines becomes significant because the laser lines are essentially shifted intact to a lower radio frequency called an intermediate frequency (IF). With a direct detection receiver, consisting of a photodetector followed by a filter, laser line shape is not a consideration since the photodetector cannot differentiate between narrow-band optical frequencies. The spectral width at the IF becomes a problem if frequency or phase detection is employed, since the line width represents a phase uncertainty. With any type of optical mixing, some form of phase or frequency detection is necessary in order to frequency or phase lock the laser carrier to the local oscillator. Hence the phase uncertainty or phase noise is a problem even for an intensity modulation laser communication system if a heterodyne or homodyne receiver is employed.

Phase noise arises from random frequency shifting of the carrier and local oscillator lasers as a result of spontaneous emission quantum noise and environmental vibrations of the laser cavity mirrors. The quantum noise is a fundamental irreducible cause of frequency fluctuations. In most practical situations the quantum noise is masked by the much stronger environmental noise. Carrier frequency fluctuations due to atmospheric turbulence are another major source of phase noise.

The signal voltage v_{IF} at the output of the filter of an optical heterodyne receiver, for perfect spatial alignment of the carrier and local oscillator, is derived in Equation 10-38 as

$$v_{IF} = G\mathscr{D}A_C A_O R_L \cos\left[(\omega_O - \omega_C)t + (\Phi_O - \Phi_C)\right] \equiv A \cos\left[\omega_{IF}t + \Phi_{IF}(t)\right] \tag{8-36}$$

where:

$$G = \text{photodetector gain}$$
$$\mathscr{D} = \text{detector conversion factor}$$
$$A_C = \text{carrier amplitude}$$
$$A_O = \text{local oscillator amplitude}$$
$$R_L = \text{optical receiver load resistance}$$
$$\omega_O = \text{local oscillator angular frequency}$$
$$\omega_C = \text{carrier angular frequency}$$
$$\Phi_O = \text{local oscillator phase angle}$$
$$\Phi_C = \text{carrier phase angle}$$

In the absence of atmospheric turbulence, A is nearly constant since both the carrier and local oscillator lasers are amplitude stabilized. The phase jitter of the IF signal may be defined as the change in phase between a time t and a time $t + \tau$ later.

$$\Delta\Phi_{IF}(t, \tau) \equiv \Phi_{IF}(t + \tau) - \Phi_{IF}(t) \tag{8-37}$$

Phase jitter has been found to be a Gaussian random variable when the phase noise is caused by quantum or environmental noise [8-6] or atmospheric turbulence [8-7].

Two quantities of interest in the analysis of phase noise are the mean square phase jitter and the power spectral density of the IF signal voltage. The mean square jitter is defined as

$$\overline{\Delta\Phi_{IF}^2(\tau)} \equiv \overline{|\Delta\Phi(t, \tau)|^2} \tag{8-38}$$

The mean square phase jitter and IF signal voltage power spectrum are related by [8-8].

$$G_{v_{IF}}(f) = A^2 \int_0^\infty \exp\left[-\tfrac{1}{2}\overline{\Delta\Phi_{IF}^2(\tau)}\right] \cos\left[(\omega - \omega_{IF})\tau\right]d\tau \tag{8-39}$$

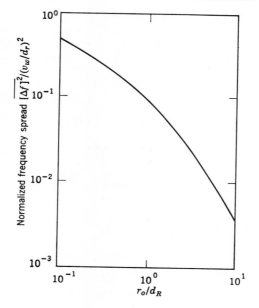

Figure 8-3

Experimental measurements indicate that the power spectrum of v_{IF} has a Gaussian shape and that the phase jitter has a quadratic time dependence $(\overline{\Delta\Phi_{IF}^2(\tau)} \sim \tau^2)$ when environmental noise is dominant [8-8]. The measurements are self-consistent with Equation 8-39. If quantum noise were dominant, the spectrum of v_{IF} would have a Lorentzian line shape and the mean square phase jitter would have a linear time jitter dependence [8-8].

Fried has derived expressions for the frequency spectrum and mean square frequency spread of v_{IF} in the presence of atmospheric turbulence [8-9]. Figure 8-3 illustrates the mean square frequency spread $\overline{(\Delta f)^2}$, as a function of wind velocity, v_w, receiver antenna diameter, d_R, and the phase coherence dimension, r_0, defined by Equation 7-11. For a given wind velocity and antenna size, as the useful mixing area of the reciever r_0 increases the frequency spread decreases.

8.5 SUMMARY OF DETECTION NOISE SOURCES

Table 8-1 lists the noise power spectral densities of the major sources of detection noise. Flicker noise and current noise are low frequency phenomena. Their effects can be bypassed by restricting the information signal

Type of Detection Noise	Power Spectral Density in Current Units
Thermal noise	$\dfrac{2kT}{R}$
Flicker noise (photoemissive detector)	$\alpha_F G^2 \dfrac{I_P^2}{f}$
Current noise (photovoltaic and photoconductive detectors)	$\alpha_c G^2 \dfrac{I_P^2}{f}$
Dark current shot noise (photoemissive and photovoltaic detectors)	$G^2 q I_D$
Photon fluctuation shot noise (photoemissive detector)	$G^2 q (I_S + I_B)$
Photon fluctuation generation-recombination noise (photovoltaic detector)	$G^2 q (I_S + I_B)$
Photon fluctuation generation-recombination noise (photoconductive detector)	$2 G^2 q (I_S + I_B)$

Table 8-1. POWER SPECTRAL DENSITIES OF MAJOR SOURCES OF OPTICAL DETECTION NOISE

bandwidth to above a low frequency cutoff of from 10 to 100 Hz, or by placing the information on a radio frequency subcarrier. The same techniques often negate the effects of intensity fluctuations of the laser carrier and background radiation, which are generally limited to low frequencies.

The shot noise and generation-recombination noise at the output of a detector. due to dark current emissions, background radiation photon fluctuations, and laser carrier photon fluctuations may be lumped together. The power spectral densities for these combined noise sources are

$$G_{i_P}(f) = G^2 q I_P \qquad \text{(photoemissive detector or} \qquad (8\text{-}40)$$
$$\text{photovoltaic detector)}$$

$$G_{i_P}(f) = 2 G^2 q I_P \qquad \text{(photoconductive detector)} \qquad (8\text{-}41)$$

where I_P is the total average detector current.

Thermal noise is a universal type of noise found in all detection systems. In radio frequency systems, thermal noise is most often the limiting factor in the detection of signals. At optical frequencies, thermal noise is also the major noise source for semiconductor detectors and photoemissive detectors without secondary gain mechanisms. Secondary electron multiplication in a photomultiplier tube usually makes the detector shot noise dominant. The dominance condition for shot and thermal noise through a zonal filter of bandwidth B is

$$2G^2qI_PBR_L > 4kTB \qquad (8\text{-}42)$$

The current gain required for shot noise dominance is thus

$$G > \sqrt{\frac{2kT}{qI_PR_L}} \qquad (8\text{-}43)$$

The current gains of semiconductor photodetectors are not presently large enough to realize shot noise limited detection in most applications. As an example, if an average signal current as large as 10^{-9} amps is to be detected (comparable to detector dark current without cooling), and if the detector

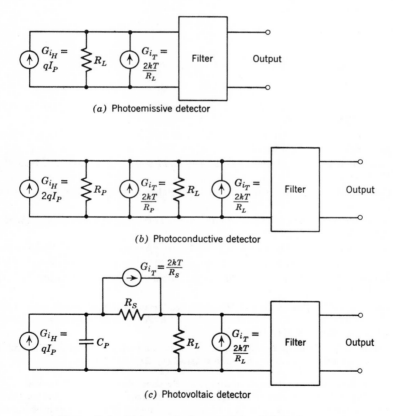

(a) Photoemissive detector

(b) Photoconductive detector

(c) Photovoltaic detector

Figure 8-4 *Noise equivalent models of photodetectors*

temperature is 300° K and the detector load resistance is 100 Ω, a detector gain of 720 is required to achieve shot noise dominance. If a photomultiplier is utilized in an optical receiver, the reception will be limited by shot noise since the current gain of a photomultiplier is about 10^6. The use of other types of detectors generally results in thermal noise limited operation.

Figure 8-4 shows the noise equivalent circuits of photoemissive, photoconductive, and photovoltaic detectors, considering only shot† and thermal noise sources. The models include the equivalent detector series and shunt resistances R_S and R_P and the shunt capacitance C_P. Following the detector is a load resistance R_L and a lossless filter.

In the models each noise source is a statically independent current generator. Hence, to determine the total noise power at the output of one of the detectors, the contributions of the individual noise sources must be combined in a root mean square sense rather than added algebraically.

REFERENCES

8-1. Johnson, J. B. "Thermal Agitation of Electricity in Conductors." *Physical Review*, **32**, 97–109, July 1928.

8-2. Nyquist, H. "Thermal Agitation of Electric Charge in Conductors." *Physical Review*, **32**, 110–113, July 1928.

8-3. Lawson, J. L. and Uhlenbeck, G. E. *Threshold Signals*. McGraw-Hill, New York, 1950.

8-4. Gadsen, M. "Some Statistical Properties of Pulses from Photomultipliers." *Applied Optics*, **4** (11), 1446–1452, Nov. 1965.

8-5. Jamieson, J. A. et al. *Infrared Physics and Engineering*. McGraw-Hill, New York, 1963.

8-6. Armstrong, J. A. "Theory of Interferometric Analysis of Laser Phase Noise." *Journal of the Optical Society of America*, **56** (8), 1024–1031, Aug. 1966.

8-7. Tatarski, V. I. *Wave Propagation in a Turbulent Medium*. McGraw-Hill, New York, 1961.

8-8. Siegman, A. E., Daino, B., and Manes, K. R. "Preliminary Measurements of Laser Short-Term Frequency Fluctuations." *IEEE Journal of Quantum Electronics*, **QE-3** (5), 180–189, May 1967

8-9. Fried, D. L. "Atmospheric Modulation Noise in an Optical Heterodyne Receiver." *IEEE Journal of Quantum Electronics*, **QE-3** (6), 213–221, June 1967.

† To simplify subsequent discussions the single term shot noise will be used for all dark current and photon current fluctuation noise sources with the stipulation that the shot noise spectral density should be multiplied by a factor of two if a photoconductive detector is employed.

chapter 9

DETECTION PROCESSES

The probability distribution for the output of a photodetector and a post detection filter is developed in this chapter for the practical operating conditions of shot and thermal noise limited detection. Knowledge of these probability distributions is vital for the design of the optical receiver electrical detector and specification of its performance.

Figure 9-1 illustrates a model for the combination of detector signal and and noise sources. The photodetector output current is the sum of the signal current, the detector current due to background radiation, and the detector dark current. Thermal noise combines with the detector current in the load resistor. This detector model applies directly to a photoemissive detector. For a photovoltaic or photoconductive detector, an additional thermal noise source, due to resistive elements in the detector, must be added at the photodetector material summation point.

9.1 PHOTODETECTOR OUTPUT PROBABILITY DENSITY

Photoelectrons are combined from the signal, background, and dark current emissions at the photodetector material summation point. As noted in Equation 8-14, the detector dark current photoelectron count is given by a Poisson distribution

$$P(U_{D,\tau} = k) = \frac{(\mu_{D,\tau})^k \exp\{-\mu_{D,\tau}\}}{k!} \qquad (9-1)$$

where $\mu_{D,\tau}$ is the average number of dark current photoelectron emissions per time period τ. The statistical distributions of laser signal and background radiation photoelectron emissions, as discussed in Section 1.2, are approximately Poisson if the detector input intensity is low, which is the operating condition of interest for an efficient laser communication system. Thus, for

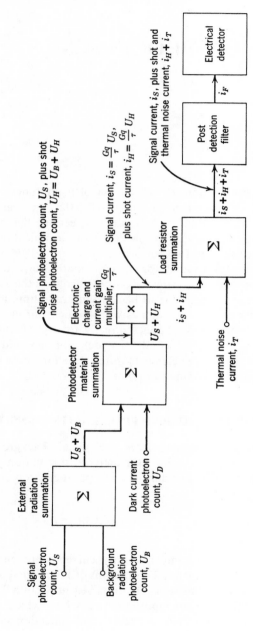

Figure 9-1   Photodetector model for combination of detector signal and noise sources

low laser and background radiation intensities, the probability distributions of photoelectron emissions due to the signal and background are

$$P(U_{S,\tau} = k) = \frac{(\mu_{S,\tau})^k \exp\{-\mu_{S,\tau}\}}{k!} \tag{9-2}$$

$$P(U_{B,\tau} = k) = \frac{(\mu_{B,\tau})^k \exp\{-\mu_{B,\tau}\}}{k!} \tag{9-3}$$

where $\mu_{S,\tau}$ and $\mu_{B,\tau}$ are the average numbers of photoelectron emissions due to laser signal and background radiation per time period τ.

The dark current, background radiation, and laser radiation photoelectron emission processes are independent; hence the probability distribution of photoelectron emissions due to all processes occurring simultaneously is also a Poisson distribution whose mean is the sum of the means of the individual processes (Appendix C). Thus, when the laser carrier and background radiation are incident upon the photodetector surface and dark current is flowing in the detector, the probability distribution of photoelectron emissions is

$$P(U_{S,\tau} + U_{H,\tau} = k) = \frac{(\mu_{S,\tau} + \mu_{H,\tau})^k \exp\{-(\mu_{S,\tau} + \mu_{H,\tau})\}}{k!} \tag{9-4}$$

where $\mu_{H,\tau} = \mu_{B,\tau} + \mu_{D,\tau}$. The corresponding probability distribution when the laser carrier is absent is

$$P(U_{H,\tau} = k) = \frac{(\mu_{H,\tau})^k \exp\{-\mu_{H,\tau}\}}{k!} \tag{9-5}$$

The probability distributions of signal plus shot noise and shot noise alone in terms of the detector currents are given by

$$P\left(i_s + i_H = \frac{kGq}{\tau}\right) = \frac{(\mu_{S,\tau} + \mu_{H,\tau})^k \exp\{-(\mu_{S,\tau} + \mu_{H,\tau})\}}{k!} \tag{9-6}$$

and

$$P\left(i_H = \frac{kGq}{\tau}\right) = \frac{(\mu_{H,\tau})^k \exp\{-\mu_{H,\tau}\}}{k!} \tag{9-7}$$

Thermal noise added at the detector output has a Gaussian distribution with zero mean and a variance proportional to the total thermal noise power. The probability density of the thermal noise current source is

$$p(i_T) = (2\pi\sigma_{i_T}^2)^{-1/2} \exp\left\{-\frac{i_T^2}{2\sigma_{i_T}^2}\right\} \tag{9-8}$$

where $\sigma_{i_T}^2 = N_T/R_L$ and N_T is the thermal noise power measured across the load resistor R_L.

In general, photodetectors are shot noise limited if the detector employs high-gain photomultiplication but otherwise are thermal noise limited. In the former case the detector output probability distributions are given by Equations 9-4 and 9-5 or 9-6 and 9-7. For thermal noise limited operation, to achieve a reasonably large signal-to-noise ratio, the laser carrier intensity must be increased to such a level that the dark current and background radiation photoelectron emissions are neglible. With large signal photoelectron counts the signal current probability distribution may be considered Gaussian [9-1]

$$P(i_S) = \left(\frac{2\pi q^2 \mu_{S,\tau}}{\tau^2}\right)^{-1/2} \exp\left\{\frac{-\left(i_S - \frac{q\mu_{S,\tau}}{\tau}\right)^2}{\frac{2q^2\mu_{S,\tau}}{\tau^2}}\right\} \quad (9\text{-}9)$$

The probability density of the total photodetector current, which is the sum of the signal and thermal noise currents, is the sum of two Gaussian variables and is therefore Gaussian (Appendix B). Hence

$$P(i_S + i_T) = \left[2\pi\left(\frac{q^2 \mu_{S,\tau}}{\tau^2} + \sigma_{i_T}^2\right)\right]^{-1/2} \exp\left\{\frac{-\left(i_P - \frac{q\mu_{S,\tau}}{\tau}\right)^2}{2\left(\frac{q^2\mu_{S,\tau}}{\tau^2} + \sigma_{i_T}^2\right)}\right\} \quad (9\text{-}10)$$

In most detection situations, the thermal noise variance, $\sigma_{i_T}^2$, will dominate the signal variance, $(q^2/\tau^2)\mu_{S,\tau}$. For this condition the probability density of the photodetector current for thermal noise limited operation is given by

$$P(i_S + i_T) = (2\pi\sigma_{i_T}^2)^{-1/2} \exp\left\{\frac{-[i_P - (q/\tau)\mu_{S,\tau}]^2}{2\sigma_{i_T}^2}\right\} \quad (9\text{-}11)$$

9.2 FILTERED PHOTODETECTOR OUTPUT PROBABILITY DENSITY

The total detector current, i_P, at the input to the post detection filter is composed of signal and shot noise current characterized by Poisson emission statistics for shot noise limited detection, or of signal and Gaussian thermal noise current for thermal noise limited detection. Effects of post detection filtering will be considered for each of these cases.

Shot Noise Limited Detection

Gilbert and Pollak [9-2] have developed a method for determining the probability distribution of a linearly filtered Poisson process. If the total detector current is composed of Poisson impulses of electronic charge q, occurring at a constant average rate of n_q per second, then the output of the post detection filter of impulse response $h(t)$ is

$$i_F = \sum_{j=1}^{k} qh(t - t_j) \tag{9-12}$$

The probability density of the filtered photodetector current is then the solution of the integral equation

$$i_F P(i_F) = n_q \int_0^{\infty} \frac{d}{di_F} \{P[i_F - qh(t)]qh(t)\} \, dt \tag{9-13}$$

Figure 9-2 illustrates the impulse response of several types of filters for which solutions to the integral equation are available. In the case of the rectangular impulse response, the photoelectrons emitted by the detector are simply counted for a given time period; probability distributions for the filter output are then given by Equations 9-4 and 9-5. From an alternate viewpoint the rectangular impulse response may be considered as characterizing an ideal integrator circuit which integrates the detector current for τ

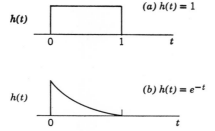

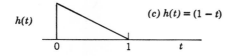

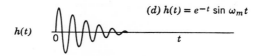

Figure 9-2   *Post detection filter impulse responses*

seconds and is then cleared for the next decision interval. In this interpretation the detector current probability distributions are given by Equations 9-6 and 9-7.

Figures 9-3, 9-4, and 9-5 give the probability distributions for the other impulse response functions of Figure 9-2. The distributions are normalized to a time scale of one second. They may be referenced to a decision interval

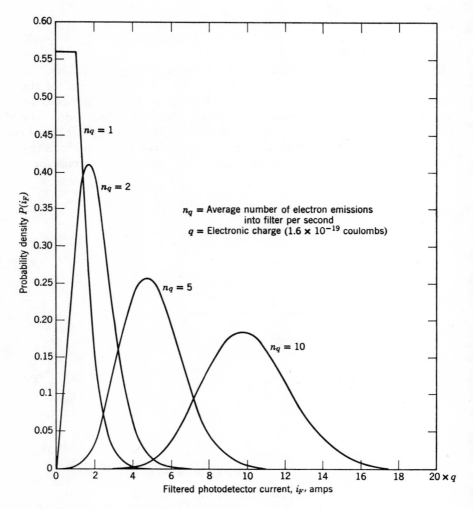

Figure 9-3 *Probability density of filtered photodetector current for filter impulse response, $h(t) = e^{-t}$*

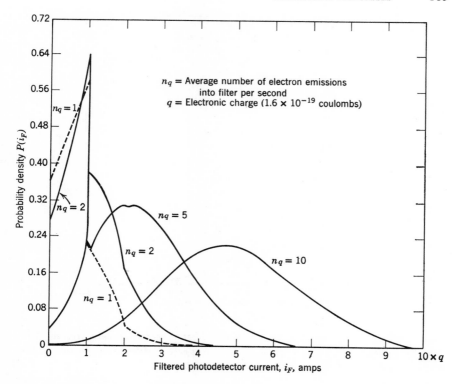

Figure 9-4 *Probability density of filtered photodetector current for filter impulse response,* $h(t) = (1 - t)$

of τ seconds by simply considering n_q to be the average number of electron emissions over τ and by multiplying the horizontal axis by $1/\tau$. In this manner it is seen that for ten or more electrons per decision interval, the probability distributions begin to approach a Gaussian distribution for the low pass filter responses of Figures 9-2b and 9-2c. The distribution of the bandpass filter output of Figure 9-2d approaches a Gausian distribution even faster [9-2].

Thermal Noise Limited Detection

Gaussian noise passing through a linear filter remains Gaussian [9-3]. If a Gaussian current, i_G, passes through a linear filter of response $h(t)$, the mean, $E[i_F]$, and autocorrelation, $R_{i_F}(\gamma)$, of the output current, i_F, are, respectively (Appendix B),

$$E[i_F] = E[i_G] \int_{-\infty}^{\infty} h(t)\, dt \qquad (9\text{-}14)$$

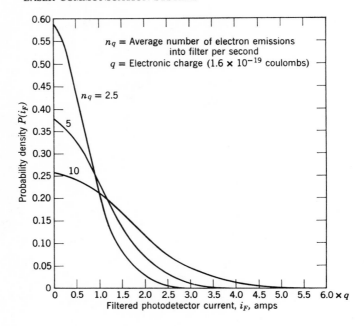

Figure 9-5 *Probability density of filtered photodetector current, i_F, for filter impulse response, $h(t) = e^{-t} \sin \omega_m t$ [9-2] (Copyright 1960, The American Telephone and Telegraph Co., reprinted by permission)*

and

$$R_{i_F}(\gamma) = \int_{-\infty}^{\infty} \int_{-\infty}^{\infty} h(\alpha)h(\beta)R_{i_F}(\gamma + \beta - \alpha)\, d\alpha\, d\beta \qquad (9\text{-}15)$$

where $E[i_G]$ and $R_{i_F}(\gamma)$ are the mean and autocorrelation of i_G, respectively. For white noise, Equation 9-15 reduces to

$$R_{i_F}(\gamma) = \sigma_{i_G}^2 \int_{-\infty}^{\infty} h(\alpha)h(\alpha - \gamma)\, d\alpha \qquad (9\text{-}16)$$

and the variance of the filter output is

$$\sigma_{i_F}^2 = R_{i_F}(0) = \sigma_{i_G}^2 \int_{-\infty}^{\infty} h^2(\alpha)\, d\alpha \qquad (9\text{-}17)$$

where $\sigma_{i_G}^2$ is the variance of i_G. Since the filtered photodetector current is Gaussian for thermal noise limited detection, its probability distribution is completely characterized by Equations 9-14 and 9-17.

9.3 SINGLE DETECTOR THRESHOLD DETECTION

In many optical detection systems a signal is determined to be present if the output of a photodetector exceeds some preestablished decision threshold. The optimum choice of this threshold is given by the likelihood ratio test of decision theory [9-4, 9-5]. This threshold value is optimum in the sense that it minimizes the probability of detection error. In the following discussion the likelihood ratio test threshold is derived for shot and thermal noise limited detection. The post detection filter for shot noise limited operation is assumed to be an ideal integrator response (Figure 9-2d) such that photoelectrons are counted over a decision interval, τ. For thermal noise limited operation, a zonal filter with a unity passband from $-B_O$ to B_O Hz is assumed.

Shot Noi   Limited Detection

The likelihood ratio test yields the decision rule for shot noise limited detection that a signal is judged to be present if

$$\frac{P(U_{S,\tau} + U_{H,\tau} = k)}{P(U_{H,\tau} = k)} = \frac{\dfrac{(\mu_{S,\tau} + \mu_{H,\tau})^k \exp\{-(\mu_{S,\tau} + \mu_{H,\tau})\}}{k!}}{\dfrac{(\mu_{H,\tau})^k \exp\{-\mu_{H,\tau}\}}{k!}} \geq \frac{1 - P(S)}{P(S)}$$

$$(9\text{-}18)$$

where $P(U_{S,\tau} + U_{H,\tau} = k)$ and $P(U_{H,\tau} = k)$ are the conditional probabilities of the photodetector emissions for signal plus shot noise and shot noise alone, and where $P(S)$ is the *a priori* probability that a signal is present. Solving for the threshold value k_T of k, for which equality exists in Equation 9-18, yields

$$k_T = \frac{\mu_{S,\tau} + \ln\left[\dfrac{1 - P(S)}{P(S)}\right]}{\ln\left[1 + \dfrac{\mu_{S,\tau}}{\mu_{H,\tau}}\right]}$$

$$(9\text{-}19)$$

The likelihood ratio threshold is, in general, not an integer. The actual decision threshold chosen, k_D, is the greatest integer value of k_T. Figures 9-6 and 9-7 give the likelihood ratio test threshold as a function of signal and shot noise photoelectron counts for pulse-code intensity modulation (PCM/IM) and pulse-position intensity modulation (PPM/IM).

During each decision interval there are: (1) a probability, P_{SN}, that the decision threshold, k_D, will be exceeded by the laser carrier and the detector

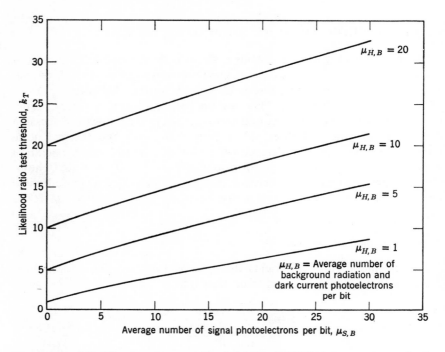

Figure 9-6 *Likelihood ratio test threshold for PCM/IM laser communication system-shot noise limited operation, $P(S) = \frac{1}{2}$*

noise and (2) a probability, P_N, that the threshold will be exceeded by noise alone. For shot noise limited operation, these detection probabilities are

$$P_{SN} = \sum_{k=k_D}^{\infty} \frac{(\mu_{S,\tau} + \mu_{H,\tau})^k \exp\{-(\mu_{S,\tau} + \mu_{H,\tau})\}}{k!} \qquad (9\text{-}20)$$

and

$$P_N = \sum_{k=k_D}^{\infty} \frac{(\mu_{H,\tau})^k \exp\{-(\mu_{H,\tau})\}}{k!} \qquad (9\text{-}21)$$

Thermal Noise Limited Detection

For thermal noise limited detection the likelihood ratio test condition for a signal to be present is

$$\frac{P(i_F \mid S)}{P(i_F \mid \bar{S})} = \frac{(2\pi\sigma_{i_T}^2)^{-1/2} \exp\left\{\dfrac{-[i_F - (q/\tau)\mu_{S,\tau}]^2}{2\sigma_{i_T}^2}\right\}}{(2\pi\sigma_{i_T}^2)^{-1/2} \exp\left\{\dfrac{-i_F^2}{2\sigma_{i_T}^2}\right\}} \geq \frac{1 - P(S)}{P(S)} \qquad (9\text{-}22)$$

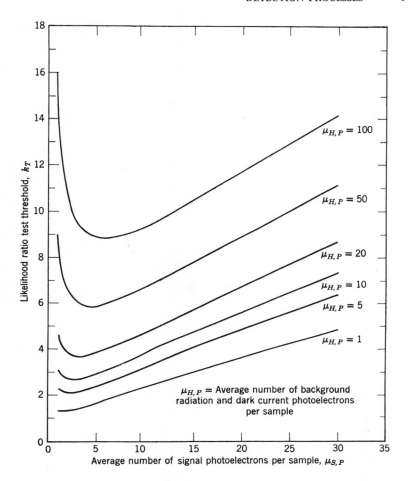

Figure 9-7 *Likelihood ratio test threshold for PPM/IM laser communication system-shot noise limited operation,* $P(S) = \frac{1}{32}$

where $P(i_F \mid S)$ and $P(i_F \mid \bar{S})$ are the conditional probability densities of the filtered detector current when the laser signal is present and absent, respectively. The threshold value of i_F, designated by $i_F{}'$ is

$$i_F{}' = \frac{q}{2\tau}\mu_{S,\tau} + \frac{\sigma_{i_T}{}^2}{(q/\tau)\mu_{S,\tau}} \ln\left[\frac{1 - P(S)}{P(S)}\right] \qquad (9\text{-}23)$$

Figures 9-8 and 9-9 show the likelihood ratio test threshold current for

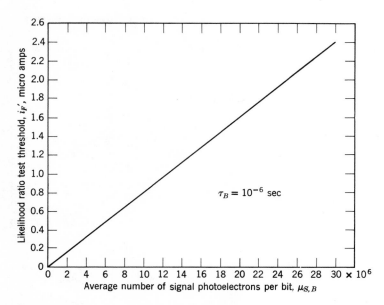

Figure 9-8 *Likelihood ratio test threshold for PCM/IM laser communication system-thermal noise limited detection,* $P(S) = \frac{1}{2}$

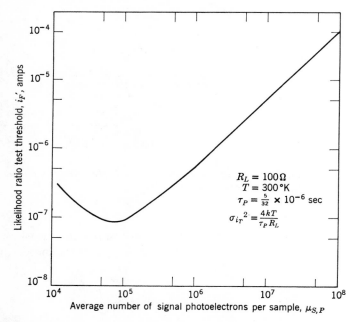

Figure 9-9 *Likelihood ratio test threshold for PPM/IM laser communication system-thermal noise limited detection,* $P(S) = \frac{1}{32}$

PCM/IM and PPM/IM, respectively. The detection probabilities for thermal noise limited detection are then

$$P_{SN} = \int_{i_F'}^{\infty} (2\pi\sigma_{i_T}{}^2)^{-1/2} \exp\left\{\frac{-[i_F - (q/\tau)\mu_{S,\tau}]^2}{2\sigma_{i_T}{}^2}\right\} di_F \qquad (9\text{-}24)$$

and

$$P_N = \int_{i_F'}^{\infty} (2\pi\sigma_{i_T}{}^2)^{-1/2} \exp\left\{\frac{-i_F{}^2}{2\sigma_{i_T}{}^2}\right\} di_F \qquad (9\text{-}25)$$

9.4 TWIN DETECTOR THRESHOLD DETECTION

Another common detection method is based upon the decision as to which of a pair of photodetectors has the largest output when the laser signal is present at only one of them. A simple method of comparing the detector outputs is to form their difference and compare the amplitude of the difference signal to a predetermined threshold network (Figure 9-10).

Shot Noise Limited Detection

Assuming the laser carrier to be entirely directed toward the X detector in Figure 9-10, the probability distributions of photoelectron counts of the photodetectors are

$$P(U_{X,\tau} = k) = \frac{(\mu_{S,\tau} + \mu_{H,\tau,R})^k \exp\{-(\mu_{S,\tau} + \mu_{H,\tau,R})\}}{k!} \qquad (9\text{-}26)$$

and

$$P(U_{Y,\tau} = k) = \frac{(\mu_{H,\tau,L})^k \exp\{-\mu_{H,\tau,L}\}}{k!} \qquad (9\text{-}27)$$

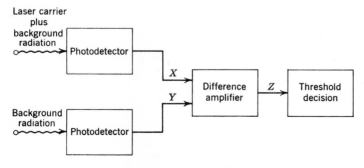

Figure 9-10   *Twin detector decision network*

where $\mu_{H,\tau,R}$ and $\mu_{H,\tau,L}$ are the average numbers of shot noise photoelectrons emitted by the X and Y detectors per time period τ. From Appendix C the probability distribution of the difference between the detector outputs can be written in terms of a modified Bessel function of order j, $I_j[\cdot]$. Thus

$$P(U_{Z,\tau} = k) = \exp\{-(\mu_{S,\tau} + 2\mu_{H,\tau})\}\left[\frac{\mu_{H,\tau,L}}{\mu_{S,\tau} + \mu_{H,\tau,R}}\right]^{-k/2}$$

$$\times I_{|k|}[2\sqrt{\mu_{H,\tau,L}(\mu_{S,\tau} + \mu_{H,\tau,R})}] \quad (9\text{-}28)$$

where $2\mu_{H,\tau} \equiv \mu_{H,\tau,R} + \mu_{H,\tau,L}$. A comparision is made by the threshold circuit as to whether the optical carrier is striking the X or Y photodetector on the basis of the value of k compared to the decision threshold. The likelihood ratio threshold k_T, is found from solution of the equation

$$\frac{\exp\{-(\mu_{S,\tau} + 2\mu_{H,\tau})\}\left[\dfrac{\mu_{H,\tau,L}}{\mu_{S,\tau} + \mu_{H,\tau,R}}\right]^{-k_T/2} \times I_{|k_T|}[2\sqrt{\mu_{H,\tau,L}(\mu_{S,\tau} + \mu_{H,\tau,R})}]}{\exp\{-(\mu_{S,\tau} + 2\mu_{H,\tau})\}\left[\dfrac{\mu_{S,\tau} + \mu_{H,\tau,L}}{\mu_{H,\tau,R}}\right]^{-k_T/2} \times I_{|k_T|}[2\sqrt{\mu_{H,\tau,R}(\mu_{S,\tau} + \mu_{H,\tau,L})}]} = \frac{P(Y)}{P(X)} \quad (9\text{-}29)$$

where $P(X)$ and $P(Y)$ are the *a priori* probabilities of the laser carrier being directed toward the X and Y photodetectors. If the shot noise is the same in each channel, i.e., $\mu_{H,\tau,R} = \mu_{H,\tau,L} \equiv \mu_{H,\tau}$, the likelihood ratio test threshold is

$$k_T = \frac{\ln\left[\dfrac{P(Y)}{P(X)}\right]}{\ln\left[1 + \dfrac{\mu_{S,\tau}}{\mu_{H,\tau}}\right]} \quad (9\text{-}30)$$

The decision threshold, k_D, is the largest integer value of k_T.

Thermal Noise Limited Detection

With the laser carrier striking the X detector only, the probability densities of the X and Y detectors are

$$P(i_X) = (2\pi\sigma_{i_{TX}}^2)^{-1/2} \exp\left\{\frac{-[i_X - (q/\tau)\mu_{S,\tau}]^2}{2\sigma_{i_{TX}}^2}\right\} \quad (9\text{-}31)$$

and

$$P(i_Y) = (2\pi\sigma_{i_{TY}}^2)^{-1/2} \exp\left\{\frac{-i_Y^2}{2\sigma_{i_{TY}}^2}\right\} \quad (9\text{-}32)$$

The probability density of the difference of the detectors is Gaussian since subtraction is a linear operation. Hence,

$$P(i_Z) = (2\pi\sigma_{i_T}^2)^{-1/2} \exp\left\{\frac{-[i_Z - (q/\tau)\mu_{S,\tau}]^2}{2\sigma_{i_T}^2}\right\} \tag{9-33}$$

where $\sigma_{i_T}^2 \equiv \sigma_{i_{TX}}^2 + \sigma_{i_{TY}}^2$. The likelihood ratio test threshold, i_Z' is then found as the solution of

$$\frac{(2\pi\sigma_{i_T}^2)^{-1/2} \exp\left\{\dfrac{-[i_Z' - (q/\tau)\mu_{S,\tau}]^2}{2\sigma_{i_T}^2}\right\}}{(2\pi\sigma_{i_T}^2)^{-1/2} \exp\left\{\dfrac{-[i_Z' + (q/\tau)\mu_{S,\tau}]^2}{2\sigma_{i_T}^2}\right\}} = \frac{P(Y)}{P(X)} \tag{9-34}$$

which is

$$i_Z' = \frac{\sigma_{i_T}^2}{(q/\tau)\mu_{S,\tau}} \ln\left[\frac{P(Y)}{P(X)}\right] \tag{9-35}$$

REFERENCES

9-1. Parzen, E. *Modern Probability Theory and Its Applications*. John Wiley, New York, 1960.

9-2. Gilbert, E. N. and Pollak, H. O. "Amplitude Distribution of Shot Noise." *Bell Systems Technical Journal*, **39**, 333–350, Mar. 1960.

9-3. Papoulis, A. *Probability, Random Variables, and Stochastic Processes*. McGraw-Hill, New York, 1965.

9-4. Reiffen, B. and Sherman, H. "An Optimum Demodulator For Poisson Processes: Photon Source Detectors." *Proceedings IEEE*, **51** (10), 1316–1320, Oct. 1963.

9-5. Curran, T. F. and Ross, M. "Optimum Detection Thresholds in Optical Communications." *Proceedings IEEE*, **53** (11), 1770–1771, Nov. 1965.

chapter 10

COMMUNICATION RECEIVERS

In Chapter 2 a basic description of optical receivers was given concerning their operation in the absence of noise. This chapter presents an analysis of direct, heterodyne, and homodyne detection optical communication receivers from a spectral viewpoint [10-1, 10-2]. Signal-to-noise ratio (SNR) expressions are derived for the receivers, considering detector shot noise and thermal noise due to resistive elements in the receivers [10-3, 10-4]. All other noise sources are assumed negligible. A table listing the SNR expressions is included at the end of the chapter.

10.1 BASEBAND DIRECT DETECTION RECEIVER

In the baseband direct detection receiver, illustrated by the detection model of Figure 10-1, signal and background radiation passing through an optical bandpass filter impinge on the surface of a photodetector. Shot noise proportional to the average detector current caused by the signal, background radiation, and dark current is generated within the detector and combined with thermal noise at the receiver output. Both the shot and thermal noise are filtered to the information signal bandwidth by the zonal low pass output filter.

Signal Power

The received laser carrier is described by an unmodulated carrier wave of instantaneous intensity

$$C(t) = A_c{}^2 \cos^2 \omega_c t \qquad (10\text{-}1)$$

where A_c is the carrier amplitude and $f_c = \omega_c/2\pi$ is the carrier frequency. The average carrier power at the detector surface, P_C, is the time average of the carrier wave over a few carrier cycles.

$$P_C = \widetilde{C(t)} = \tfrac{1}{2}A_c{}^2 \qquad (10\text{-}2)$$

Since the optical carrier is unmodulated, the instantaneous photodetector

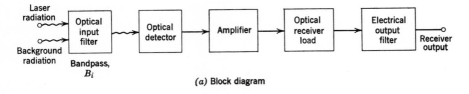

(a) Block diagram

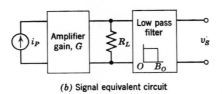

(b) Signal equivalent circuit

Figure 10-1   *Baseband direct detection optical receiver*

current, i_P, and the average photodetector current, I_P, are identical and, as noted in Equation 1-31, equal to

$$i_P = I_P = \mathscr{D}P_C \qquad (10\text{-}3)$$

where $\mathscr{D}$ is the detector intensity-to-current conversion factor.† The instantaneous receiver output signal voltage, v_s, is then equal to

$$v_S = Gi_PR_L = G\mathscr{D}P_CR_L \qquad (10\text{-}4)$$

where G is the photodetector current gain and R_L is the optical receiver load resistance. The average signal power at the receiver output is defined to be the time average of v_s^2 over a time period comparable to the reciprocal of the output filter bandwidth, B_O. Thus, for the unmodulated carrier, the signal power is

$$S = \frac{v_S^2}{R_L} = (G\mathscr{D}P_C)^2 R_L \qquad (10\text{-}5)$$

This measure of signal power is a peak value since any intensity modulation reduces the carrier intensity. As noted later by Equation 10-18, if the carrier were 100% intensity modulated by a sine wave the signal power would be one-eighth of that given by Equation 10-5.

Noise Power

If the output filter is a zonal low pass filter with unity transmission over a frequency range of $-B_O$ to B_O Hz and zero elsewhere, the shot noise power as given by the Schottky formula of Table 8-1 is

$$N_H = 2qG^2(I_P + I_B + I_D)B_OR_L \qquad (10\text{-}6)$$

† In this chapter $\mathscr{D} \equiv \eta q/hf_c$ is the value assumed for the detector conversion factor.

where I_D is the average detector dark current and where the average detector current due to the background radiation power, P_B, incident upon the detector is†

$$I_B = \mathcal{D}P_B \tag{10-7}$$

By the Johnson formula of Table 8-1, the thermal noise power is

$$N_T = 4kTB_O \tag{10-8}$$

Figure 10-2 illustrates the signal and noise spectra in the direct detection optical receiver for an arbitrary signal power spectrum.

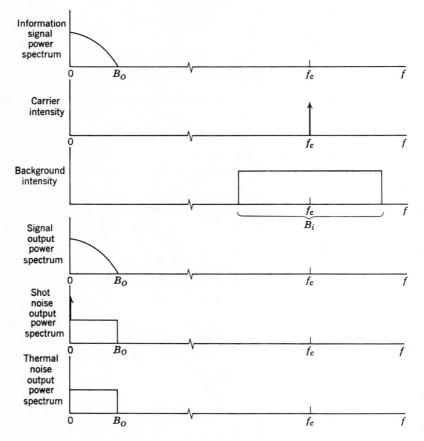

Figure 10-2   *Signal and noise spectra of baseband direct detection optical receiver*

† If a polarizer is placed before the detector, the expressions for P_B given in Table 6-2 should be reduced by one-half. In this situation the carrier must be properly polarized.

Signal-to-Noise Ratio

A common performance criterion of communication theory is the signal-to-noise ratio (SNR) defined as the ratio of the average signal power to average noise power at some point in the receiver chain. For optical communications the SNR has fault as a performance measure because the detector noise is a function of carrier power. Chapter 11 illustrates situations in which a high SNR does not guarantee a low probability of detection error. However, the SNR measure can be useful if limitations are placed upon the relative amounts of carrier and background radiation entering the receiver.

The SNR at the input to the first detector of a communication receiver (the photodetector for an optical receiver) is called the carrier-to-noise ratio (CNR). In a radio frequency system, the noise input to the first detector is usually caused by thermal noise from the antenna and associated preamplifier. An optical receiver has the distinct advantage that optical antennas do not introduce thermal noise; the CNR reduces to the ratio of the laser carrier radiation power, P_C, to the background radiation power, P_B.

$$\frac{C}{N} \equiv \frac{P_C}{P_B}$$

The SNR is defined as the ratio of the average signal power, as defined by Equation 10-5, to the shot and thermal noise power, as given by Equations 10-6 and 10-8. Thus, the SNR at the receiver output, in terms of the average carrier and background radiation power incident upon the photodetector, is

$$\frac{S}{N} = \frac{\left[\frac{G\eta q}{hf_c}\right]^2 R_L P_C^2}{2qB_O G^2 \left\{\frac{\eta q}{hf_c}[P_C + P_B] + I_D\right\} R_L + 4kTB_O} \tag{10-9}$$

If the photodetector gain is unity, the thermal noise is usually much greater than the shot noise, and the SNR becomes

$$\frac{S}{N} = \left[\frac{\eta q}{hf_c}\right]^2 \frac{R_L P_C^2}{4kTB_O} \tag{10-10}$$

On the other hand, for G large, shot noise predominates and the SNR reduces to

$$\frac{S}{N} = \left[\frac{\eta}{hf_c}\right]^2 \frac{q}{2B_O} \frac{P_C^2}{\left\{\frac{\eta q}{hf_c}[P_C + P_B] + I_D\right\}} \tag{10-11}$$

For the case in which the dark current is made negligible by cooling the detector, the SNR may be written in terms of the CNR, as

$$\frac{S}{N} = \frac{\eta}{2hf_cB_O} \left[\frac{P_C}{1 + \dfrac{1}{(C/N)}} \right] \tag{10-12}$$

The direct detection receiver SNR is maximum for carrier shot noise limited operation when the background radiation is zero or, equivalently, when the CNR is infinite. In this situation the SNR becomes

$$\frac{S}{N} = \frac{\eta P_C}{2hf_cB_O} \tag{10-13}$$

Equation 10-13 represents the limiting capability of a direct detection optical receiver in which the only noise is the uncertainty in the number of signal photoelectrons emitted by the detector.

10.2 SUBCARRIER DIRECT DETECTION RECEIVER

In a subcarrier direct detection optical communication system, an optical carrier is intensity modulated by a high frequency radio subcarrier wave, which is itself modulated by an information signal [10-5]. Figure 10-3 contains a block diagram of a subcarrier direct detection optical receiver. The output of the optical detector is a sine wave of the subcarrier frequency, plus detector shot noise, filtered by a zonal electrical bandpass filter. An electrical detector, to be discussed later, removes the information signal from the subcarrier and completes the demodulation process.

Subcarrier Signal Power

The received laser carrier is described by a carrier wave which is intensity modulated by a radio frequency subcarrier sine wave:

$$C(t) = \tfrac{1}{2}[1 + A_{SC} \cos(\omega_{SC}t + \Phi_{SC})]A_c{}^2 \cos^2 \omega_c t \tag{10-14}$$

where:

$$A_{SC} = \text{subcarrier amplitude } (|A_{SC}| \leq 1)$$

$$f_{SC} = \omega_{SC}/2\pi = \text{subcarrier frequency}$$

$$\Phi_{SC} = \text{subcarrier phase}$$

This subcarrier wave may be amplitude, frequency, or phase modulated.

The instantaneous detector current due to the subcarrier is then

$$i_P = \mathscr{D}\widetilde{C(t)} = \frac{\mathscr{D}}{4} A_c{}^2[1 + A_{SC} \cos(\omega_{SC}t + \Phi_{SC})] \tag{10-15}$$

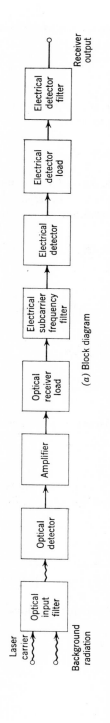

(a) Block diagram

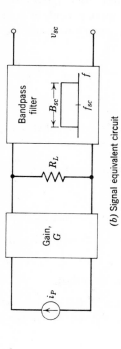

(b) Signal equivalent circuit

Figure 10-3 Subcarrier direct detection optical receiver

with an average or direct current value that can be related to the unmodulated average carrier power, P_C.

$$I_P = \frac{\mathcal{D}A_c^2}{4} = \frac{\mathcal{D}P_C}{2} \tag{10-16}$$

The resulting instantaneous voltage generated across the optical receiver load resistance, R_L, and passed by the bandpass filter is

$$v_{SC} = \frac{G\mathcal{D}P_C A_{SC} R_L}{2} \cos\left[\omega_{SC}t + \Phi_{SC}\right] \tag{10-17}$$

An effective peak subcarrier signal power may be defined as the time average, over the subcarrier wave period, of the instantaneous subcarrier signal power with the subcarrier wave amplitude set to its maximum value, i.e., $A_{SC} = 1$. Thus,

$$[S]_{SC} = \frac{\overline{[v_{SC}]^2}}{R_L} = \frac{G^2\mathcal{D}^2 P_C^2 R_L}{8} \tag{10-18}$$

Subcarrier Noise Power

At the output of the zonal subcarrier filter the expressions for the shot and thermal noise power are

$$[N_H]_{SC} = 2qG^2\left\{\mathcal{D}\frac{P_C}{2} + \mathcal{D}P_B + I_D\right\}B_{SC}R_L \tag{10-19}$$

and

$$[N_T]_{SC} = 4kTB_{SC} \tag{10-20}$$

Figure 10-4 illustrates the signal and noise spectra at the subcarrier filter output.

Subcarrier Signal-to-Noise Ratio

Forming the ratio of the expressions for the subcarrier signal power to the subcarrier shot and thermal noise power gives the SNR at the subcarrier filter output

$$\left[\frac{S}{N}\right]_{SC} = \frac{\left[G\frac{\eta q}{hf_c}\right]^2 P_C^2 R_L}{16G^2 q\left\{\frac{\eta q}{hf_c}\left[\frac{P_C}{2} + P_B\right] + I_D\right\}B_{SC}R_L + 32kTB_{SC}} \tag{10-21}$$

The subcarrier SNR achieves its highest value when the optical receiver is signal shot noise limited. In this case

$$\left[\frac{S}{N}\right]_{SC} = \frac{\eta P_C}{8hf_c B_{SC}} \tag{10-22}$$

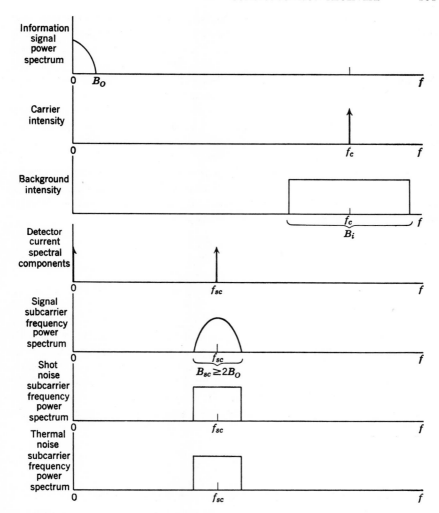

Figure 10-4   *Signal and noise spectra of subcarrier output of subcarrier direct detection optical receiver*

Subcarrier Second Detection

A second detection operation must now be performed by the electrical detector to obtain the information signal from the subcarrier. Since the subcarrier filter output voltage is linearly proportional to the subcarrier wave, any linear AM, FM, or PM radio receiver provides proper demodulation.

From Figure 9-5 it is seen that shot noise passed through a bandpass filter is closely approximated by a Gaussian distribution. Thermal noise due to the optical receiver load resistor, R_L, is also Gaussian (Section 8-1) at the subcarrier filter output. Thus, with accuracy sufficient for most applications, the subcarrier filter output may be considered to be a sine wave of the subcarrier frequency plus narrow-band Gaussian noise.

For amplitude modulation of the subcarrier, a synchronous or linear envelope detector may be employed for second detection. A synchronous detector gives an output SNR equal to twice the subcarrier filter output SNR when the detector input is composed of a sine wave plus narrow-band Gaussian noise [10-6]. Under the same conditions, if the subcarrier SNR is reasonably large, the envelope detector output SNR is also twice the subcarrier SNR [10-6]. For a synchronous detector or an envelope detector with $[S/N]_{SC}$ large, the receiver output SNR becomes

$$\frac{S}{N} = 2\left[\frac{S}{N}\right]_{SC} \tag{10-23}$$

Doppler shifting of the carrier frequency does not cause the subcarrier frequency at the detector output to shift since the photodetector output is insensitive to the carrier frequency over the operating range of the detector. The doppler shift of the subcarrier is less than that of the carrier by the ratio of the carrier to subcarrier frequencies, and in many instances the subcarrier doppler shift is negligible. Hence, if the subcarrier oscillator at the transmitter is relatively stable, the subcarrier filter bandwidth, B_{SC}, can be ideally set at twice the electrical detector output filter bandwidth, B_O. With $B_{SC} = 2B_O$, for signal shot noise limited detection and amplitude modulation, the output SNR is

$$\frac{S}{N} = \frac{\eta P_C}{8hf_c B_O} \tag{10-24}$$

Comparison with Equation 10-13 indicates that the SNR for the subcarrier receiver is one-fourth that which could be obtained by using direct detection and direct modulation of the carrier.

If the subcarrier is frequency modulated, a limiter-discriminator radio receiver may be used for second detection. The subcarrier filter band-width, B_{SC}, for FM, as given by Equation 2-20, is usually set at

$$B_{SC} = 2f_d\left[1 + \frac{1}{M_{FM}}\right] \tag{10-25}$$

where f_d is the frequency deviation and M_{FM} is the FM modulation index.

With a limiter-discriminator receiver the output SNR, for $[S/N]_{SC}$ large, and for a sine wave plus narrow-band Gaussian noise input, is [10-6]

$$\frac{S}{N} = 3(M_{FM})^2 \frac{B_{SC}}{2B_O} \left[\frac{S}{N}\right]_{SC} \tag{10-26}$$

Again, for signal shot noise limited detection the receiver output SNR for frequency modulation of the subcarrier is

$$\frac{S}{N} = \frac{3}{16} (M_{FM})^2 \frac{\eta P_C}{h f_c B_O} \tag{10-27}$$

10.3 HETERODYNE DETECTION RECEIVER

In a heterodyne detection optical receiver, as shown in Figure 10-5, the incoming laser carrier is combined with a reference wave from a local oscillator on the photodetector surface [10-7 to 10-11]. The optical mixing of the signal and local oscillator waves produces an intermediate frequency (IF) difference signal carrying the laser carrier modulation. This IF signal passes through a zonal bandpass filter to an electrical second detector for final demodulation. The difference frequency is continuously monitored at the input of the electrical detector, and the local oscillator frequency is varied to keep the IF center frequency constant. Frequency control of the local oscillator is necessary to correct for frequency drifting of the laser carrier and local oscillator and also to compensate for doppler shifts in the case of space communications.

The principal advantages of heterodyne operation are the relative ease of amplification at an intermediate frequency and the fact that the local oscillator power may be set to swamp out both the thermal noise and the shot noise caused by all sources other than the local oscillator itself to improve the SNR.

IF Signal Power

Figure 10-6 illustrates the spatial combination of the local oscillator plane wave and the laser carrier wave which has been collimated by the receiver optical antenna [10-12, 10-13]. For simplicity the photodetector surface is assumed to be square with dimension d, and the laser carrier is assumed to be misaligned by the angle ψ in only one coordinate. In the analysis, let

$$E(t) = A_c \cos \left[\omega_c t + \Phi_c - \frac{\omega_c x}{v_x} \right] \tag{10-28}$$

represent the electric field amplitude of the received laser carrier with average power $P_C = \frac{1}{2}A_c^2$, and let

$$L(t) = A_o \cos (\omega_o t + \Phi_o) \tag{10-29}$$

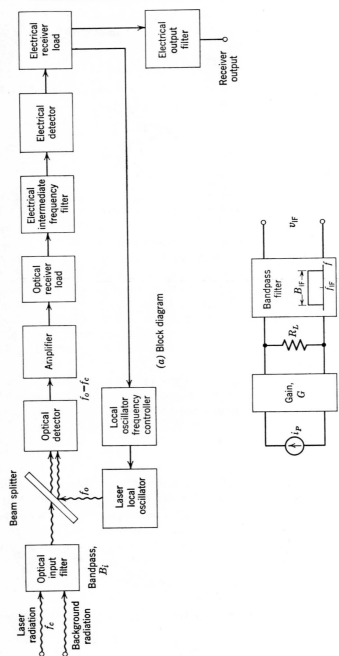

Figure 10-5 *Heterodyne detection optical receiver*

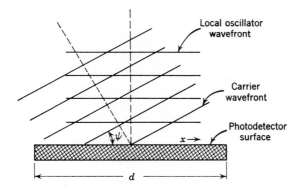

Figure 10-6 *Spatial combination of carrier and local oscillator for collimating receiver optical antenna*

be the local oscillator electric field with average power $P_O = \frac{1}{2}A_o^2$ where v_x is the laser carrier wave velocity along the detector surface. The carrier and local oscillator instantaneous *amplitudes* combine at the photodetector surface to yield an instantaneous detector input signal intensity equal to

$$C(t) = [E(t) + L(t)]^2 = \left[A_c \cos\left(\omega_c t + \Phi_c - \frac{\omega_c x}{v_x} \right) + A_o \cos(\omega_o t + \Phi_o) \right]^2$$

$$(10\text{-}30)$$

The resultant instantaneous carrier and local oscillator current at the photodetector output is the time average and spatial integral of the light intensity over the detector surface

$$i_P = \frac{\mathcal{D}}{d^2} \int_{-d/2}^{+d/2} \int_{-d/2}^{+d/2} \left\{ A_c^2 \cos^2\left(\omega_c t + \Phi_c - \frac{\omega_c x}{v_x} \right) + A_o^2 \cos^2(\omega_o t + \Phi_o) \right.$$

$$+ A_c A_o \cos\left[(\omega_o - \omega_c)t + (\Phi_o - \Phi_c) + \frac{\omega_c x}{v_x} \right]$$

$$\left. + A_c A_o \cos\left[(\omega_o + \omega_c)t + (\Phi_o + \Phi_c) - \frac{\omega_c x}{v_x} \right] \right\} dx\, dy \quad (10\text{-}31)$$

The average or direct current photodetector current due to the carrier and local oscillator is

$$I_P = \frac{\mathcal{D}}{2} (A_c^2 + A_o^2) = \mathcal{D}(P_C + P_O) \quad (10\text{-}32)$$

In Equation 10-31 the difference frequency term is unaffected by the short time average, while the other terms are filtered out by the IF filter. At the IF filter output, the instantaneous IF signal voltage is then

$$v_{IF} = \frac{G\mathscr{D}R_L}{d} \int_{-d/2}^{+d/2} \int_{-d/2}^{+d/2} \left\{ A_c A_o \cos \left[(\omega_o - \omega_c)t + (\Phi_o - \Phi_c) + \frac{\omega_c x}{v_x} \right] \right\} dx\, dy$$

$$(10\text{-}33)$$

Performing the integration yields

$$v_{IF} = G\mathscr{D}A_c A_o R_L \cos \left[(\omega_o - \omega_c)t + (\Phi_o - \Phi_c) \right] \frac{\sin (\omega_c d/2v_x)}{(\omega_c d/2v_x)} \quad (10\text{-}34)$$

From Figure 10-6 the wave velocity along the detector surface is

$$v_x = \frac{c}{\sin \psi} \tag{10-35}$$

where c is the velocity of light. Hence, the amplitude of the IF voltage is dependent upon the misalignment angle, ψ. To keep the signal phase cancellation due to misalignment to 10% or less, the term $(\omega_c d/2v_x)$ in Equation 10-34 must be 0.8 radians or less. From this requirement it is found that

$$\psi \lesssim \frac{\lambda}{4d} \tag{10-36}$$

As an example, at a wavelength of 10^{-4} cm and for a detector dimension of 1 cm, the spatial misalignment angle, ψ, and hence the receiver pointing accuracy must be less than 25 microradians.

If the laser carrier wave is focused to a diffraction limited spot by a focusing type of optical antenna (Figure 10-7) instead of being collimated, the beam

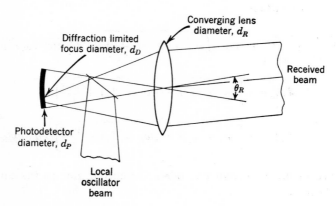

Figure 10-7 *Spatial combination of carrier and local oscillator for focusing receiver optical antenna*

misalignment angle is determined by the receiver field of view [10-13]. In this configuration the local oscillator beam is made divergent to illuminate the photodetector surface uniformly. From Equation 1-13 the receiver field of view, θ_R, may be expressed as

$$\theta_R = 2.44\lambda \frac{d_P}{d_D d_R} \qquad (10\text{-}37)$$

where d_P is the photodetector diameter, d_D is the diffraction limited spot diameter, and d_R is the receiver antenna diameter.

For the previous example with a photodetector of 1 cm diameter and an optical antenna of 10 cm diameter, giving a diffraction limited spot of 0.01 cm diameter at a transmission wavelength of 10^{-4} cm, the receiver field of view is approximately 2.5 milliradians. Use of a focusing type of receiver antenna rather than a collimating antenna therefore provides a relaxation in the antenna pointing requirements by a factor of 100.

It should be noted that with the focusing antenna, photomixing only occurs over the diffraction limited spot on the photosurface. The portion of the local oscillator beam striking other regions of the detector contributes to the shot noise without providing a useful signal and, hence, lowers the SNR. This problem can be eliminated by the use of an image dissector phototube in which only the photocurrent from the focal spot is utilized [10-14, 10-15].

If the receiver is aligned perfectly, the instantaneous voltage at the output of the IF filter referenced to the optical receiver load resistance, R_L, will be

$$v_{IF} = G\mathcal{D}A_c A_o R_L \cos\left[(\omega_o - \omega_c)t + (\Phi_o - \Phi_c)\right] \qquad (10\text{-}38)$$

The resulting effective signal power at the IF output is the time average over the intermediate frequency wave period of the instantaneous IF power. Thus

$$[S]_{IF} = \frac{\overline{[v_{IF}]^2}}{R_L} = 2G^2\mathcal{D}^2 P_o P_c R_L \qquad (10\text{-}39)$$

IF Noise Power

The shot and thermal noise powers at the output of the zonal IF filter are

$$[N_H]_{IF} = 2G^2 q[\mathcal{D}(P_C + P_O + P_B) + I_D]B_{IF}R_L \qquad (10\text{-}40)$$

and

$$[N_T]_{IF} = 4kTB_{IF} \qquad (10\text{-}41)$$

Figure 10-8 illustrates the signal and noise spectra at the output of the IF filter.

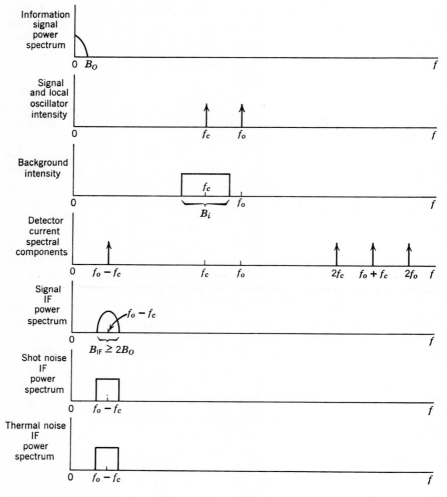

Figure 10-8   *Signal and noise spectra of IF output of heterodyne detection optical receivers*

IF Signal-to-Noise Ratio

The SNR at the output of the IF filter is then

$$\left[\frac{S}{N}\right]_{\mathrm{IF}} = \frac{\left[\dfrac{G\eta q}{hf_c}\right]^2 P_O P_C R_L}{G^2 q B_{\mathrm{IF}} \left\{\dfrac{\eta q}{hf_c}\left[P_C + P_O + P_B\right] + I_D\right\} R_L + 2\,kTB_{\mathrm{IF}}} \qquad (10\text{-}42)$$

If the local oscillator power is large, all signal, background, and dark current shot noise effects, plus the thermal noise, will be swamped out by the local oscillator shot noise. The SNR then reduces to

$$\left[\frac{S}{N}\right]_{\text{IF}} = \frac{\eta P_C}{h f_c B_{\text{IF}}} \tag{10-43}$$

Comparison of Equation 10-43 with Equation 10-22 indicates that the SNR at the electrical second detector input is at least eight times higher for the heterodyne system than for the subcarrier direct detection system if the band-pass filter bandwidths are the same. If the detector thermal noise or dark current is appreciable, or if the background radiation power level is large, the SNR advantage of the heterodyne over the subcarrier system becomes even greater.

IF Second Detection

An electrical second detection operation must now be performed to obtain the information signal from the IF subcarrier. Since the IF filter output voltage, Equation 10-38, is linearly proportional to the carrier electric field amplitude, A_c, a linear rectifier or synchronous detector provides proper second detection for *amplitude* modulation of the carrier. Square-law second detection results in proper demodulation for an *intensity* modulated carrier. Frequency demodulation can be performed by a limiter-discriminator detector. When the local oscillator amplitude is large, the emission statistics of the photodetector may be considered Gaussian.

For a linear envelope detector with $[S/N]_{\text{IF}}$ large, or for a synchronous detector, the output SNR for a sine wave plus narrow-band Gaussian noise input is

$$\frac{S}{N} = 2\left[\frac{S}{N}\right]_{\text{IF}} \tag{10-44}$$

Assuming that the IF filter bandwidth is set at $B_{\text{IF}} = 2B_O$, the output SNR for a strong local oscillator becomes

$$\frac{S}{N} = \frac{\eta P_C}{h f_c B_O} \tag{10-45}$$

The SNR is eight times that obtained with a subcarrier AM system under signal shot noise limited conditions.

For a sine wave plus narrow-band Gaussian noise input the output SNR for a square-law radio detector is equal to one-half the IF SNR [10-6]

$$\frac{S}{N} = \frac{1}{2}\left[\frac{S}{N}\right]_{\text{IF}} \tag{10-46}$$

which for a strong local oscillator for $B_{IF} = 2B_O$ becomes

$$\frac{S}{N} = \frac{\eta P_C}{4hf_cB_O} \tag{10-47}$$

Comparision with Equation 10-13 indicates that for intensity modulation the SNR of a heterodyne detection receiver is one-half that of a signal shot noise limited direct detection receiver.

A limiter-discriminator radio receiver produces an output SNR equal to

$$\frac{S}{N} = 3(M_{FM})^2 \frac{B_{IF}}{2B_O} \left[\frac{S}{N}\right]_{IF} \tag{10-48}$$

For a strong local oscillator the SNR reduces to

$$\frac{S}{N} = \frac{3}{2}(M_{FM})^2 \frac{\eta P_C}{hf_cB_O} \tag{10-49}$$

With frequency modulation the heterodyne detection receiver SNR is eight times as large as the SNR of the subcarrier direct detection receiver operating under signal shot noise limited conditions.

When comparing the SNR of the IF second detector output of a heterodyne detection receiver to the SNR of the final output of a direct detection receiver, the ratio B_{IF}/B_O and the amount of noise due to causes other than signal shot noise must be considered. If B_{IF} is made much larger than $2B_O$ to accomodate doppler shifting of the carrier and laser frequency instability, the heterodyne detection receiver SNR will be degraded. On the other hand, if thermal noise or shot noise due to background radiation and dark current emissions in a direct detection receiver is appreciable, its SNR will be much smaller than the values given by Equations 10-13, 10-24, or 10-27.

10.4 HOMODYNE DETECTION RECEIVER

In the optical homodyne receiver shown in Figure 10-9, the local oscillator reference is set at the same frequency and phase as the laser carrier before optical mixing on the photodetector surface. The resultant photodetector output contains the information signal at the baseband. A local oscillator phase controller monitors the receiver output signal and adjusts the phase of the oscillator to match the carrier.

Signal Power

For perfect spatial alignment of the carrier and local oscillator, their electric fields may be represented as

$$E(t) = A_c \cos(\omega_c t + \Phi_c) \tag{10-50}$$

$$L(t) = A_o \cos(\omega_c t + \Phi_o) \tag{10-51}$$

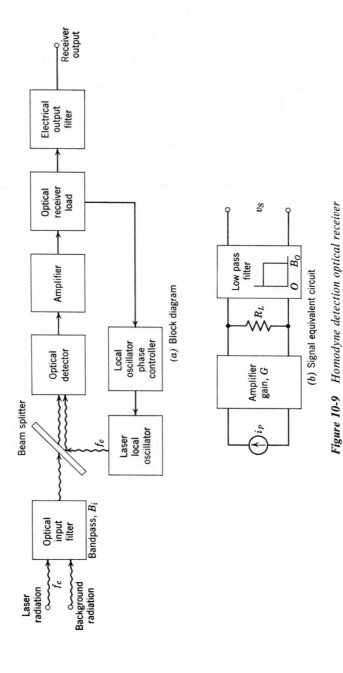

Figure 10-9 Homodyne detection optical receiver

The instantaneous photodetector current is then

$$i_P = \mathscr{D}\{\tfrac{1}{2}A_c{}^2 + \tfrac{1}{2}A_o{}^2 + A_cA_o \overbrace{\cos{(\Phi_o - \Phi_c)}}$$

$$+ A_cA_o \overbrace{\cos{[2\omega_c t + (\Phi_o + \Phi_c)]}}\} \quad (10\text{-}52)$$

The phase difference term of Equation 10-52 is unaffected by the short time average while the double frequency term is removed by the filter, leaving the useful signal voltage

$$v_S = \mathscr{D}A_cA_oR_L \cos{(\Phi_o - \Phi_c)} \quad (10\text{-}53)$$

The direct current carrier term of Equation 10-52 does not contribute significantly to the output signal, since for proper operation the local oscillator power is set much larger than the carrier power ($A_cA_o \gg \tfrac{1}{2}A_c{}^2$).

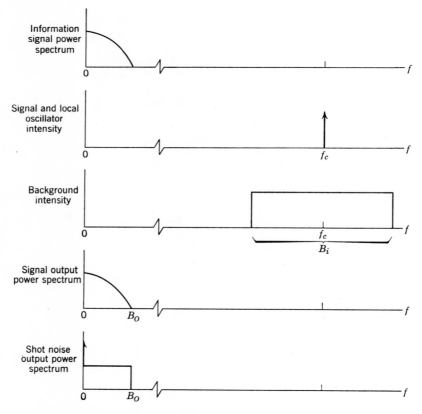

Figure 10-10   *Signal and noise spectra of homodyne optical receiver*

For amplitude modulation the carrier and local oscillator are phase locked so that $\Phi_o = \Phi_c$; for phase modulation Φ_o is constant and Φ_c is proportional to the time-varying information signal. In either case the peak signal power is

$$S = 4G^2\mathscr{D}^2P_OP_CR_L \qquad (10\text{-}54)$$

Noise Power

If the local oscillator power is much larger than the carrier power, the shot noise power becomes

$$N_H = 2qG^2B_O[\mathscr{D}(P_O + P_B) + I_D]R_L \qquad (10\text{-}55)$$

and the thermal noise power is

$$N_T = 4kTB_O \qquad (10\text{-}56)$$

Figure 10-10 illustrates the signal and noise spectra in a homodyne receiver.

Signal-to-Noise Ratio

The signal-to-noise ratio of the homodyne receiver is then

$$\frac{S}{N} = \frac{2\left(G\dfrac{\eta q}{hf_c}\right)^2 P_OP_CR_L}{G^2q\left[\dfrac{\eta q}{hf_c}(P_O + P_B) + I_D\right]B_OR_L + 2kTB_O} \qquad (10\text{-}57)$$

If the local oscillator power is large, the thermal noise and shot noise due to the carrier, background radiation, and dark current are swamped out, yielding

$$\frac{S}{N} = \frac{2\eta P_C}{hf_cB_O} \qquad (10\text{-}58)$$

The SNR of a homodyne receiver is thus four times as large as the SNR of a signal shot noise limited direct detection receiver.

Table 10-1 lists expressions for the signal-to-noise ratios of optical receivers. Relations are given for the general case in which both thermal and shot noise are present in the optical receiver. Expressions also are developed for digital communication systems for thermal and shot noise limited operation in terms of the average number of carrier, background radiation, and dark current photoelectrons emitted by the receiver detector in a bit period.

System	General Cases, Thermal and Shot Noise†	Thermal Noise Limited Operation ($G = 1$), Digital Data, $B_O = 1/\tau_B$	Shot Noise Limited Operation, Digital Data, $B_O = 1/\tau_B$
Direct detection receiver SNR	$\dfrac{\left[\dfrac{G\eta q}{hf_c}\right]^2 P_C{}^2 R_L}{2qG^2\left\{\dfrac{\eta q}{hf_c}[P_C + P_B] + I_D\right\}B_O R_L + 4kTB_O}$	$\dfrac{q^2 R_L (\mu_{S,B})^2}{4kT\tau_B}$	$\dfrac{(\mu_{S,B})^2}{2[\mu_{S,B} + \mu_{B,B} + \mu_{D,B}]}$
Subcarrier direct detection receiver, subcarrier filter SNR $B_{SC} = 2B_O$	$\dfrac{\left[\dfrac{G\eta q}{hf_c}\right]^2 P_C{}^2 R_L}{32G^2 q\left\{\dfrac{\eta q}{hf_c}\left[\dfrac{P_C}{2} + P_B\right] + I_D\right\}B_O R_L + 64kTB_O}$	$\dfrac{q^2 R_L (\mu_{S,B})^2}{64kT\tau_B}$	$\dfrac{(\mu_{S,B})^2}{32\left[\dfrac{\mu_{S,B}}{2} + \mu_{B,B} + \mu_{D,B}\right]}$
Heterodyne detection receiver, intermediate frequency filter SNR $B_{IF} = 2B_O$	$\dfrac{\eta P_C}{2hf_c B_O}$	—	$\dfrac{\mu_{S,B}}{2}$
Homodyne detection receiver SNR	$\dfrac{2\eta P_c}{hf_c B_O}$	—	$2\mu_{S,B}$

† See Appendix D for definition of terms.

Table 10-1. Optical Receiver Signal-to-Noise Ratio Expressions

REFERENCES

10-1. Kerr, J. R. "Microwave-Bandwidth Optical Receiver Systems." *Proceedings IEEE*, **55** (10), 1686–1700, Oct. 1967.

10-2. Fried, D. L. and Seidman, J. B. "Heterodyne and Photo-Counting Receivers for Optical Communications." *Applied Optics*, **6** (2), 245–250, Feb. 1967.

10-3. Oliver, B. M. "Signal-to-Noise Ratios in Photoelectric Mixing." *Proceedings IRE Correspondence*, **49** (12), 1960–1961, Dec. 1961.

10-4. Haus, H. A., Townes, C. H., and Oliver, B. M. "Comments on Noise in Photo-electric Mixing." *Proceedings IRE Correspondence*, **50** (6), 1544–1545, June 1962.

10-5. Ward, J. H. and Shechet, M. L. "Optical Subcarrier Communications." *Electrical Communication*, **42** (2), 247–260, 1967.

10-6. Panter, P. F. *Modulation, Noise, and Spectral Analysis.* McGraw-Hill, New York, 1965.

10-7. Jacobs, S. "The Optical Heterodyne." *Electronics*, **36** (28), 29, July 12, 1963.

10-8. Siegman, A. E., Harris, S. E., and McMurtry, B. J. "Optical Heterodyning and Optical Demodulation at Microwave Frequencies." *Optical Masers.* J. Fox, ed., John Wiley, New York, 1963.

10-9. Warden, M. P. "Experimental Study of the Theory of Optical Superheterodyne Reception." *Proceedings IEE*, **113** (6), 997–1004, July 1966.

10-10. Jacobs, S. F. and Rabinowitz, P. J. "Optical Heterodyning with a C. W. Gaseous Laser." *Proceedings 1963 Quantum Electronics Conference*, Columbia Univ. Press, pp. 481–487, 1964.

10-11. Lucy, R. F. et al. "Optical Superheterodyne Receiver." *Applied Optics*, **6** (8), 1333–1342, Aug. 1967.

10-12. Corcoran, V. J. "Directional Characteristics in Optical Heterodyne Detection Processes." *Journal Applied Physics*, **36**, 1819–1825. June 1965.

10-13. Siegman, A. E. "The Antenna Properties of Optical Heterodyne Receivers." *Proceedings IEEE*, **54** (10), 1350–1356, Oct. 1966.

10-14. Read, W. S. and Fried, D. L. "Optical Heterodyning with Non-critical Angular Alignment." *Proceedings IEEE Correspondence*, **51** (12), 1787–1788, Dec. 1963.

10-15. Read, W. S. and Turner, R. G. "Tracking Heterodyne Detection." *Applied Optics*, **4** (12), 1570–1573, Dec. 1965.

chapter 11

BASEBAND PULSE AND DIGITAL LASER COMMUNICATION SYSTEMS

Operational descriptions of several types of baseband pulse and digital laser communication systems are presented in the following sections. The communication systems considered are: a PCM intensity modulation system and a PCM polarization modulation system, each employing direct detection; a PCM phase modulation system using a homodyne detection receiver; and a quantized PPM intensity modulation system with a direct detection receiver. The probability of detection error is derived for the systems for shot and thermal noise limited operating conditions. Perfect time synchronization of the data is assumed.

11.1 PCM INTENSITY MODULATION SYSTEM

In the PCM intensity modulation system, bits are coded as the presence or absence of the laser carrier—a "one" bit is represented by the carrier and a "zero" bit by no carrier. A block diagram of a direct detection PCM/IM system is shown in Figure 11-1. In this system a continuous wave laser is intensity modulated to either pass the laser beam or inhibit it, as determined by the data. At the receiver the optical carrier is collected and passed through an optical filter to the photodetector. The photodetector output is filtered to the information bandwidth, and a bit decision is made as to whether the detector output exceeds a decision threshold during a bit period.

During each bit interval there is a probability, $P_{SN}{}^B$, that the decision threshold will be exceeded by the carrier signal and detector noise; there is also a probability, $P_N{}^B$, that the threshold will be exceeded by noise alone. The probability of making an incorrect bit-detection decision, $P_e{}^B$, is the probability that the signal and noise do not exceed the threshold given that the carrier signal is present, plus the probability that noise alone exceeds the

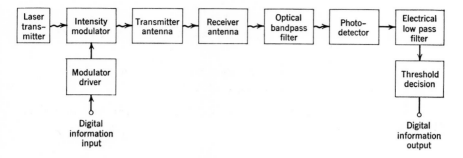

Figure 11-1 *PCM intensity modulation laser communication system*

threshold given that the carrier is not transmitted. In terms of the detection probabilities,

$$P_e^B = p(1 - P_{SN}^B) + (1 - p)P_N^B \tag{11-1}$$

where p is the probability of transmitting the laser carrier. For a system transmitting "one" and "zero" bits with equal likelihood,

$$P_e^B = \tfrac{1}{2}(1 - P_{SN}^B + P_N^B) \tag{11-2}$$

PCM/IM Shot Noise Limited Detection

For shot noise limited operation the threshold decision circuit ideally consists of an electron counter. If photomultiplication is included in the detector, the *number* of current pulses is counted during each bit period. In both cases the probability distribution of emissions at the output of the low pass filter is assumed Poisson. The total number of counts, k, during a bit period is compared to a preestablished decision threshold, k_D. If k is equal to or greater than k_D, a "one" bit is assumed to have been transmitted; if k is less than k_D, a "zero" bit decision is made. The detection probabilities for the ideal electron counter, as given in Equations 9-20 and 9-21, are

$$P_{SN}^B = \sum_{k=k_D}^{\infty} \frac{(\mu_{S,B} + \mu_{H,B})^k \exp\{-(\mu_{S,B} + \mu_{H,B})\}}{k!} \tag{11-3}$$

and

$$P_N^B = \sum_{k=k_D}^{\infty} \frac{(\mu_{H,B})^k \exp\{-\mu_{H,B}\}}{k!} \tag{11-4}$$

where:

$\mu_{S,B}$ = average number of photoelectrons emitted per bit due to the laser signal

$\mu_{H,B}$ = average number of photoelectrons emitted per bit due to background radiation and dark current†

Substitution of the above equations in Equation 11-2 yields the probability of detection error for $p = \frac{1}{2}$.

$$P_e^B = \frac{1}{2}\left\{1 - \sum_{k=k_D}^{\infty} \frac{\exp\{-\mu_{H,B}\}}{k!}\left[(\mu_{S,B} + \mu_{H,B})^k \exp\{-\mu_{S,B}\} - (\mu_{H,B})^k\right]\right\}$$

(11-5)

The value of the detection threshold which minimizes the probability of error is given by the likelihood ratio test. For $p = \frac{1}{2}$ the likelihood ratio test threshold, as shown in Figure 9-6, is

$$k_T = \frac{\mu_{S,B}}{\ln\left[1 + (\mu_{S,B}/\mu_{H,B})\right]}$$

(11-6)

The electron count decision threshold, k_D, is the largest integer value of k_T. Figure 11-2 is a plot of the probability of detection error for the decision threshold based upon the likelihood ratio test. The cusps in the curves are due to integer changes in the detection threshold as $\mu_{S,B}$ increases.

PCM/IM Thermal Noise Limited Detection

For thermal noise limited operation, to achieve a reasonably low error rate the carrier power must be increased to a relatively high level. Under this circumstance the detector current pulses merge into a continuous current signal, and it is the *amplitude* of the filtered detector current, i_F, that is compared to the current decision threshold, i_F'. Gaussian statistics are used to describe the current fluctuations at the low pass filter output. The detection probabilities for such a thermal noise limited receiver, as given in Equations 9-24 and 9-25, are

$$P_{SN}^B = \int_{i_F'}^{\infty} (2\pi\sigma_{i_T}^2)^{-1/2} \exp\left\{-\frac{\left[i_F - \frac{q}{\tau_B}\mu_{S,B}\right]^2}{2\sigma_{i_T}^2}\right\} di_F$$

(11-7)

$$P_N^B = \int_{i_F'}^{\infty} (2\pi\sigma_{i_T}^2)^{-1/2} \exp\left\{-\frac{i_F^2}{2\sigma_{i_T}^2}\right\} di_F$$

(11-8)

† If a polarizer preceeds the detector, $\mu_{H,B} = \eta P_B \tau_B / 2hf_c + I_D \tau_B / q$ where τ_B is the bit period and P_B is the total unpolarized background radiation power given in Table 6-2.

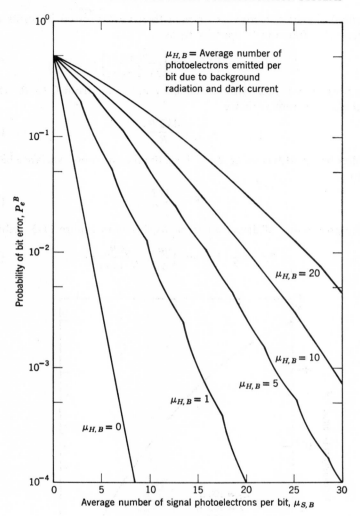

Figure 11-2 *Probability of detection error for PCM/IM direct detection laser communication system-shot noise limited operation*

where τ_B is the bit period and $\sigma_{i_T}^2$ is the thermal noise current variance. The detection threshold for $p = \frac{1}{2}$, as given in Equation 9-23, is

$$i_F' = \frac{q}{2\tau_B}\mu_{S,B} \qquad (11\text{-}9)$$

The probability of detection error may then be expressed in terms of the Gaussian error function (Appendix B) as

$$P_e{}^B = \frac{1}{2}\left\{1 - \text{erf}\left[\frac{q\mu_{S,B}}{2\sqrt{2}\ \tau_B\sigma_{i_T}}\right]\right\} \tag{11-10}$$

For an ideal zonal filter with a frequency passband from 0 to B_O Hz, the thermal noise current variance is

$$\sigma_{i_T}{}^2 = \frac{4kTB_O}{R_L} \tag{11-11}$$

If the filter bandwidth is set at $B_O = 1/\tau_B$, the thermal noise variance becomes equal to

$$\sigma_{i_T}{}^2 = \frac{4kT}{\tau_B R_L} \tag{11-12}$$

Then the probability of detection error, as shown in Figure 11-3, reduces to

$$P_e{}^B = \frac{1}{2}\left\{1 - \text{erf}\left(\frac{q\mu_{S,B}}{4\sqrt{2}}\sqrt{\frac{R_L}{kT\tau_B}}\right)\right\} \tag{11-13}$$

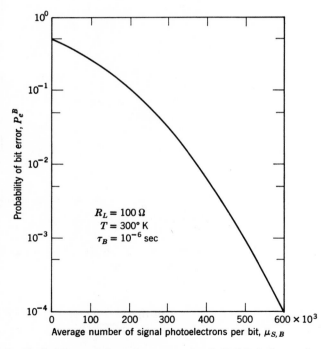

Figure 11-3 *Probability of detection error for PCM/IM direct detection laser communication system-thermal noise limited operation*

It is readily apparent from the horizontal scales of Figures 11-2 and 11-3 that the shot noise limited system is considerably more efficient than the thermal noise limited system.

11.2 PCM POLARIZATION MODULATION SYSTEM

The pulse-code modulation, polarization modulation system (PCM/PL) [11-1 to 11-4] is based upon the representation of bits as light of either right or left circular polarization. An operational block diagram of this system is shown in Figure 11-4. A polarization modulator converts the continuous wave laser beam into right or left circular polarization. At the receiver the beam passes through an optical filter and is then converted to horizontal or vertical linear polarization by a quarter wave, phase-retardation plate. The linear polarization components are spatially separated so that operationally a laser carrier of right circular polarization strikes the upper photodetector, and a carrier of left circular polarization strikes the lower photodetector. A mark or space bit decision is based upon which detector output is larger over a bit period or, equivalently, upon whether the difference between the detector output signals integrated over a bit period is positive or negative. Figure 11-5 illustrates the static operation of the polarization modulation system with full and partial modulator phase retardation.

Polarization States

The polarization modulation system may be analyzed by the polarization matrix (Appendix A) of the laser beam as it passes through the polarizing components of the communication system [11-2, 11-4]. The laser is plane polarized at a $45°$ angle with respect to the modulator axis which defines the coordinate reference system. Thus, the polarization matrix of the laser is

$$\mathscr{L} = \sqrt{\frac{P_C}{2}} \begin{bmatrix} 1 \\ 1 \end{bmatrix} \tag{11-14}$$

where P_C is the average carrier power. The modulator introduces a phase shift of Γ between the X and Y components of the incoming beam. Its polarization matrix operator is

$$M_M = \begin{bmatrix} e^{i\Gamma/2} & 0 \\ 0 & e^{-i\Gamma/2} \end{bmatrix} \tag{11-15}$$

The quarter wave plate introduces a phase shift of $\pi/2$ radians between the X and Y components and has an operator matrix

$$M_Q = \begin{bmatrix} e^{i\pi/4} & 0 \\ 0 & e^{-i\pi/4} \end{bmatrix} \tag{11-16}$$

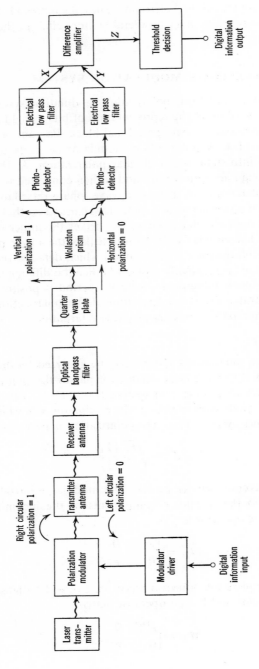

Figure 11-4 PCM polarization modulation laser communication system

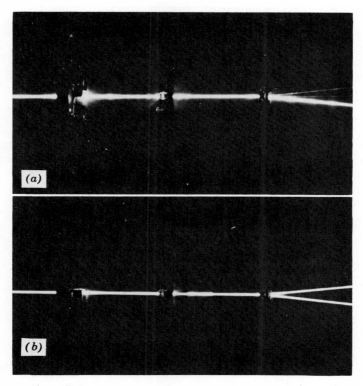

Figure 11-5 *Static operation of PCM/PL laser communication system. (a) Right circular polarization. (b) elliptical polarization. Components, left to right: Modulator, quarter wave plate, and Wollaston prism*

Finally, the right and left components of the Wollaston prism have operator matrices

$$M_{WR} = \frac{1}{2}\begin{bmatrix} 1 & -1 \\ -1 & 1 \end{bmatrix} \tag{11-17}$$

$$M_{WL} = \frac{1}{2}\begin{bmatrix} 1 & 1 \\ 1 & 1 \end{bmatrix} \tag{11-18}$$

The polarization matrices of the outputs of the modulator, quarter wave plate, and Wollaston prism are listed in Table 11-1 for an arbitrary modulator phase shift Γ and for $\Gamma = \pm\pi/2$. If the modulator does not produce a full positive or negative quarter wave phase retardation, the output of the modulator will be elliptically polarized rather than circularly polarized. This

Modulator Phase Retardation	$+\Gamma$	$-\Gamma$	$+\dfrac{\pi}{2}$	$-\dfrac{\pi}{2}$
Laser output: $\mathscr{L}$	$\sqrt{\dfrac{P_c}{2}}\begin{bmatrix}1\\1\end{bmatrix}$	$\sqrt{\dfrac{P_c}{2}}\begin{bmatrix}1\\1\end{bmatrix}$	$\sqrt{\dfrac{P_c}{2}}\begin{bmatrix}1\\1\end{bmatrix}$	$\sqrt{\dfrac{P_c}{2}}\begin{bmatrix}1\\1\end{bmatrix}$
Modulator output: $M_M\mathscr{L}$	$\sqrt{\dfrac{P_c}{2}}\begin{bmatrix}e^{i\Gamma/2}\\e^{-i\Gamma/2}\end{bmatrix}$	$\sqrt{\dfrac{P_c}{2}}\begin{bmatrix}e^{-i\Gamma/2}\\e^{i\Gamma/2}\end{bmatrix}$	$\sqrt{\dfrac{P_c}{2}}\,e^{i\pi/4}\begin{bmatrix}1\\-i\end{bmatrix}$	$\sqrt{\dfrac{P_c}{2}}\,e^{-i\pi/4}\begin{bmatrix}1\\i\end{bmatrix}$
Quarter wave plate output: $M_Q M_M\mathscr{L}$	$\sqrt{\dfrac{P_c}{2}}\begin{bmatrix}e^{i(\Gamma/2+\pi/4)}\\e^{-i(\Gamma/2+\pi/4)}\end{bmatrix}$	$\sqrt{\dfrac{P_c}{2}}\begin{bmatrix}e^{-i(\Gamma/2-\pi/4)}\\e^{i(\Gamma/2+\pi/4)}\end{bmatrix}$	$\sqrt{\dfrac{P_c}{2}}\begin{bmatrix}1\\-1\end{bmatrix}$	$\sqrt{\dfrac{P_c}{2}}\begin{bmatrix}1\\1\end{bmatrix}$
Wollaston prism right channel output: $M_{WR}M_Q M_M\mathscr{L}$	$\sqrt{\dfrac{P_c}{2}}\cos\left(\dfrac{\Gamma}{2}+\dfrac{\pi}{4}\right)\begin{bmatrix}1\\1\end{bmatrix}$	$\sqrt{\dfrac{P_c}{2}}\cos\left(\dfrac{\Gamma}{2}-\dfrac{\pi}{4}\right)\begin{bmatrix}1\\1\end{bmatrix}$	$\sqrt{\dfrac{P_c}{2}}\begin{bmatrix}0\\0\end{bmatrix}$	$\sqrt{\dfrac{P_c}{2}}\begin{bmatrix}1\\1\end{bmatrix}$
Wollaston prism left channel output: $M_{WL}M_Q M_M\mathscr{L}$	$\sqrt{\dfrac{P_c}{2}}\,i\sin\left(\dfrac{\Gamma}{2}+\dfrac{\pi}{4}\right)\begin{bmatrix}1\\-1\end{bmatrix}$	$\sqrt{\dfrac{P_c}{2}}\,i\sin\left(\dfrac{\Gamma}{2}-\dfrac{\pi}{4}\right)\begin{bmatrix}1\\1\end{bmatrix}$	$\sqrt{\dfrac{P_c}{2}}\begin{bmatrix}1\\-1\end{bmatrix}$	$\sqrt{\dfrac{P_c}{2}}\begin{bmatrix}0\\0\end{bmatrix}$

Table 11-1. POLARIZATION MATRIX DESCRIPTION OF LASER RADIATION

elliptically polarized light will produce a high-intensity linear polarization in one direction and a low-intensity linear polarization orthogonal to it. Thus, the Wollaston prism output will contain both right and left components, and both photodetectors will receive light indications. The result is that the useful transmitted signal is the difference between the high- and low-intensity components. Insufficient modulation thus degrades the received signal power. The effect is exactly analogous to less than 100% amplitude modulation in a radio frequency communication system. However, both components create shot noise so that the SNR is degraded.

Background radiation which is considered unpolarized may be represented by a two-dimensional coherency matrix

$$J = \frac{P_B}{2} \begin{bmatrix} 1 & 0 \\ 0 & 1 \end{bmatrix} \tag{11-19}$$

Table 11-2 illustrates the coherency of the background radiation as it passes through the system. The input power of the unpolarized background radiation splits equally between the channels.

PCM/PL Signal-to-Noise Ratio

Each channel of the PCM/PL receiver consists of a direct detection receiver. If the transmitted carrier is predominately right circularly polarized, the detector signal current outputs averaged over a bit period are

$$I_{S,R} = G\mathcal{D}M_{\mathrm{PL}}P_C \tag{11-20}$$

$$I_{S,L} = G\mathcal{D}(1 - M_{\mathrm{PL}})P_C \tag{11-21}$$

where M_{PL} ($0 \leq M_{\mathrm{PL}} \leq 1$) is the polarization modulation index which indicates the proportion of the carrier power in the right circular polarization state to the total carrier power, P_C, G is the detector current gain, and $\mathcal{D}$ is the detector conversion factor. The signal power at the output of the difference amplifier referred to the optical receiver load resistance, R_L, is then

$$S = (I_{S,R} - I_{S,L})^2 R_L = (2M_{\mathrm{PL}} - 1)^2 (G\mathcal{D}P_C)^2 R_L \tag{11-22}$$

The average background radiation currents at the detector outputs are

$$I_{B,R} = G\mathcal{D}P_{B,R} \tag{11-23}$$

$$I_{B,L} = G\mathcal{D}P_{B,L} \tag{11-24}$$

where:

$P_{B,R}$ = background radiation power incident upon right channel photodetector

$P_{B,L}$ = background radiation power incident upon left channel photodetector

Radiation Description	Coherency Matrix, J
Unpolarized background radiation receiver input	$\begin{bmatrix} \dfrac{P_B}{2} & 0 \\ 0 & \dfrac{P_B}{2} \end{bmatrix}$
Quarter wave plate output	$\begin{bmatrix} \dfrac{P_B}{2} & 0 \\ 0 & \dfrac{P_B}{2} \end{bmatrix}$
Wollaston prism right channel output	$\begin{bmatrix} \dfrac{P_B}{4} & -\dfrac{P_B}{4} \\ -\dfrac{P_B}{4} & \dfrac{P_B}{4} \end{bmatrix}$
Wollaston prism left channel output	$\begin{bmatrix} \dfrac{P_B}{4} & \dfrac{P_B}{4} \\ \dfrac{P_B}{4} & \dfrac{P_B}{4} \end{bmatrix}$

Table 11-2. COHERENCY MATRIX DESCRIPTION OF BACKGROUND RADIATION

Then the shot noise power at each detector output is given by the Schottky formula,

$$N_{H,R} = 2GqB_O(I_{S,R} + I_{B,R} + I_{D,R})R_L \tag{11-25}$$

$$N_{H,L} = 2GqB_O(I_{S,L} + I_{B,L} + I_{D,L})R_L \tag{11-26}$$

where B_O is the output filter bandwidth and $I_{D,R}$ and $I_{D,L}$ are the dark currents of the right and left channel photodetectors. These shot noise powers add linearly to give a total shot noise power at the difference amplifier output of

$$N_H = 2G^2qB_O(\mathscr{D}P_C + \mathscr{D}P_B + I_{D,R} + I_{D,L})R_L \tag{11-27}$$

where $P_B = P_{B,R} + P_{B,L}$. For thermal noise contributions only at the difference amplifier, the thermal noise power is

$$N_T = 4kTB_O \tag{11-28}$$

If the detector dark currents are equal, the SNR may be written as

$$\frac{S}{N} = \frac{(2M_{\text{PL}} - 1)^2(G\mathscr{D}P_C)^2R_L}{2G^2qB_O(\mathscr{D}P_C + \mathscr{D}P_B + 2I_D)R_L + 4kTB_O} \tag{11-29}$$

From Equation 11-29 it is seen that the SNR is reduced considerably if 100% ($M_{PL} = 1$) polarization modulation is not achieved. Also, the SNR is lower for the twin photodetector receiver than for the single photodetector (see Equation 10-9) due to the dark current and background radiation shot noise contribution from each detector.

PCM/PL Shot Noise Limited Detection

The probability of detection error may be derived by letting

U_X = right channel electron count

U_Y = left channel electron count

U_Z = difference electron count ($U_Z \equiv U_X - U_Y$)

If the carrier is modulated to transmit a "one" bit with 100% polarization modulation, the probability distributions of the X and Y channels are

$$P[U_X = k] = \frac{(\mu_{S,B} + \mu_{H,B})^k \exp\{-(\mu_{S,B} + \mu_{H,B})\}}{k!} \qquad (11\text{-}30)$$

and

$$P[U_Y = k] = \frac{(\mu_{H,B})^k \exp\{-\mu_{H,B}\}}{k!} \qquad (11\text{-}31)$$

where $\mu_{H,B} = \eta P_B \tau_B / 2hf_c + I_D \tau_B / q$ and τ_B is the bit period.

A detection error occurs when $U_Y > U_X$ with probability one or when $U_Z = 0$ with probability one-half, assuming "ones" and "zeros" are transmitted with equal likelihood. The probability of detection error is then, by symmetry of the channels, equal to

$$P_e^B = 1 + \tfrac{1}{2}P[U_Z = 0] - \sum_{j=0}^{\infty} P[U_Z = j] \qquad (11\text{-}32)$$

From Equation C-20 the probability distribution of U_Z for $j \geq 0$ is

$$P[U_Z = j] = \exp\{-(\mu_{S,B} + 2\mu_{H,B})\}\left[\frac{\mu_{S,B} + \mu_{H,B}}{\mu_{H,B}}\right]^{j/2}$$
$$\times I_j[2\sqrt{\mu_{H,B}(\mu_{S,B} + \mu_{H,B})}] \quad (11\text{-}33)$$

where $I_j[\cdot]$ is the modified Bessel function of order j. The probability of detection error may then be written as

$$P_e^B = 1 - \exp\{-(\mu_{S,B} + 2\mu_{H,B})\}\left[\tfrac{1}{2}I_0[2\sqrt{\mu_{H,B}(\mu_{S,B} + \mu_{H,B})}]\right.$$

$$\left. + \sum_{j=1}^{\infty}\left[\frac{\mu_{S,B} + \mu_{H,B}}{\mu_{H,B}}\right]^{j/2} I_j[2\sqrt{\mu_{H,B}(\mu_{S,B} + \mu_{H,B})}]\right] \quad (11\text{-}34)$$

The probability of error expression can be reduced to

$$P_e{}^B = \tfrac{1}{2}\{1 + Q[a, b] - Q[b, a]\} \tag{11-35}$$

where

$$a \equiv \sqrt{2\mu_{H,B}}$$

$$b \equiv \sqrt{2(\mu_{S,B} + \mu_{H,B})}$$

and

$$Q[a, b] \equiv \int_b^\infty \exp\left\{\frac{-(a^2 + x^2)}{2}\right\} I_0(ax) x \, dx \tag{11-36}$$

is Marcum's Q function of radar theory [11-5, 11-6]. The Q function has been tabulated and a recursive computational method has been developed [11-7, 11-8]. Figure 11-6 is a plot of the probability of detection error as a function of the number of signal and shot noise photoelectron counts.

PCM/PL Thermal Noise Limited Detection

For the case of thermal noise limited detection, let

$$i_X = \text{right channel current}$$

$$i_Y = \text{left channel current}$$

$$i_Z = \text{difference current } (i_Z \equiv i_X - i_Y)$$

The probability densities of i_X and i_Y are Gaussian as is the probability density of the difference current, i_Z. From Equation 9-33 the conditional densities of i_Z, given that the carrier is directed to the X and Y detectors, are

$$P(i_Z \mid S_X) = (2\pi\sigma_{i_T}{}^2)^{-1/2} \exp\left\{\frac{-\left(i_Z - \dfrac{q}{\tau_B}\mu_{S,B}\right)^2}{2\sigma_{i_T}{}^2}\right\} \tag{11-37}$$

and

$$P(i_Z \mid S_Y) = (2\pi\sigma_{i_T}{}^2)^{-1/2} \exp\left\{\frac{\left(i_Z + \dfrac{q}{\tau_B}\mu_{S,B}\right)^2}{2\sigma_{i_T}{}^2}\right\} \tag{11-38}$$

From Equation 9-35 the decision threshold is zero for "ones" and "zeros" transmitted with equal likelihood. The probability of detection error is then

$$P_e{}^B = \tfrac{1}{2}\int_{-\infty}^0 P(i_Z \mid S_X) \, di_Z + \tfrac{1}{2}\int_0^\infty P(i_Z \mid S_Y) \, di_Z \tag{11-39}$$

In terms of the Gaussian error function

$$P_e{}^B = \frac{1}{2}\left[1 - \text{erf}\left(\frac{q\mu_{S,B}}{2\tau_B\sigma_{i_T}}\right)\right] \tag{11-40}$$

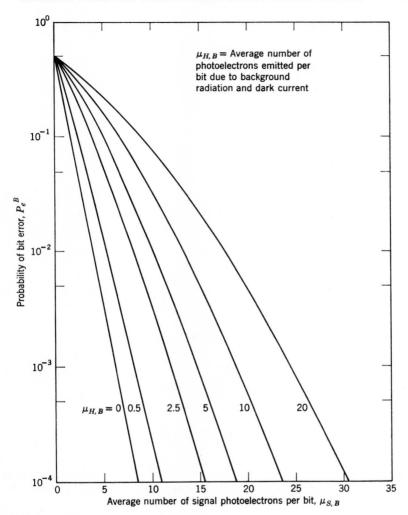

Figure 11-6 *Probability of detection error for PCM/PL direct detection laser communication system-shot noise limited operation*

Upon substitution from Equation 11-12 for the thermal noise current variance, $\sigma_{i_T}^2$, the probability of error expression reduces to

$$P_e^B = \frac{1}{2}\left[1 - \mathrm{erf}\left(\frac{q\mu_{S,B}}{4}\sqrt{\frac{R_L}{kT\tau_B}}\right)\right] \tag{11-41}$$

Figure 11-7 is a plot of the probability of detection error for thermal noise limited operation.

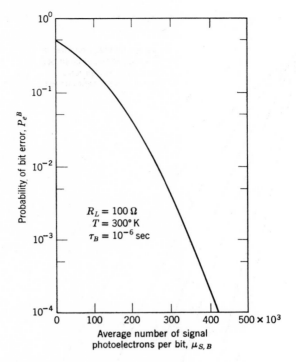

Figure 11-7 *Probability of detection error for PCM/PL direct detection laser communication system-thermal noise limited operation*

11.3 PCM PHASE MODULATION SYSTEM

The phase of the optical carrier is set at 0 or π radians, with respect to some arbitrary reference, to represent bits in the PCM phase modulation system. A block diagram of a PCM/PM system employing homodyne detection is shown in Figure 11-8.

From Equation 10-53, the signal voltage, v_S, of an optical homodyne receiver is

$$v_S = G\mathscr{D}A_cA_oR_L \cos(\Phi_o - \Phi_c) \tag{11-42}$$

where:

$\quad G$ = detector gain

$\quad \mathscr{D}$ = detector conversion factor (nq/hf_c)

$\quad A_c$ = carrier amplitude

A_o = local oscillator amplitude

R_L = optical receiver load resistance

Φ_c = carrier phase angle

Φ_o = local oscillator phase angle

The phase of the local oscillator is set such that the signal voltage is $v_S = G\mathscr{D}A_cA_oR_L$ for a "one" bit and $v_S = -G\mathscr{D}A_cA_oR_L$ for a "zero" bit. Under the conditions of a high-power local oscillator the variance of the output voltage of the homodyne receiver, from Equation 10-55, is

$$\sigma_{v_H}{}^2 = [N_H]R_L = qG^2\mathscr{D}B_oA_o{}^2R_L{}^2 \tag{11-43}$$

If the local oscillator amplitude is large the photodetector emissions may be assumed to obey Gaussian statistics. Then the conditional probability distributions of the receiver output voltage, v_S, are

$$P(v_S \mid S_1) = (2\pi\sigma_{v_H}{}^2)^{-1/2} \exp\left\{\frac{-(v_S - GDA_cA_oR_L)^2}{2\sigma_{v_H}{}^2}\right\} \tag{11-44}$$

and

$$P(v_S \mid S_0) = (2\pi\sigma_{v_H}{}^2)^{-1/2} \exp\left\{\frac{-(v_S + GDA_cA_oR_L)^2}{2\sigma_{v_H}{}^2}\right\} \tag{11-45}$$

For "ones" and "zeroes" transmitted with equal likelihood, the decision threshold is zero. A "one" is judged to have been transmitted if $v_S > 0$, and a "zero" if $v_S < 0$. The probability of detection error is then

$$P_e{}^B = \tfrac{1}{2}\int_{-\infty}^{0} P(v_S \mid S_1)\, dv_S + \tfrac{1}{2}\int_{0}^{\infty} P(v_S \mid S_0)\, dv_S \tag{11-46}$$

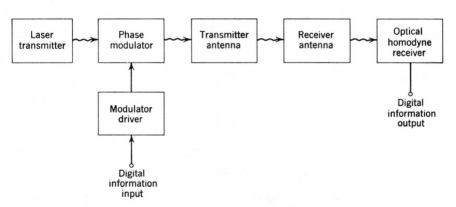

Figure 11-8   *Homodyne PCM phase modulation laser communication system*

which can be written in terms of the Gaussian error function as

$$P_e{}^B = \frac{1}{2}\left[1 - \mathrm{erf}\,\sqrt{\frac{(GDA_cA_oR_L)^2}{2\sigma_{v_H}{}^2}}\right] \tag{11-47}$$

Substituting for the shot noise current variance from Equation 11-43 yields

$$P_e{}^B = \frac{1}{2}\left[1 - \mathrm{erf}\,\sqrt{\frac{\eta A_c{}^2}{2hf_cB_O}}\right] \tag{11-48}$$

Since the average carrier power is $P_C = \frac{1}{2}A_c{}^2$, the probability of detection error can be written as

$$P_e{}^B = \tfrac{1}{2}[1 - \mathrm{erf}\,\sqrt{\mu_{S,B}}] \tag{11-49}$$

for $B_O = 1/\tau_B$. Figure 11-9 illustrates the probability of detection error as a function of the signal photoelectron count.

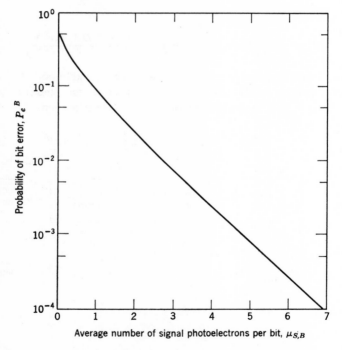

Figure 11-9 *Probability of detection error for PCM/PM homodyne detection laser communication system*

11.4 PPM INTENSITY MODULATION SYSTEM

In the optical pulse-position modulation (PPM) communication system, information is conveyed by varying the time of occurrence of a single optical pulse within a data sample-time period. The sample period is divided into discrete time intervals, and the initiation point of the optical pulse is set into one of the time slots.

A block diagram of a PPM direct detection laser communication system is shown in Figure 11-10. A laser is pulse-intensity modulated by a modulation driver fed by the output of a PPM data coder. The receiver collects the transmitted signal and passes it through an optical filter to a photodetector whose electrical output is sensed to determine if an optical pulse has been transmitted during a time interval. The time occurrence of the pulse, with respect to the beginning of the data sample period, determines the demodulated information signal.

To evaluate the probability of making an incorrect detection decision, it is necessary to postulate a set of decoding rules. For this system, threshold detection is assumed.† The sampling period, τ_P, is divided into K equal duration time slots, and a detector threshold count is established. The received signal is said to occur in the time slot for which the photodetector

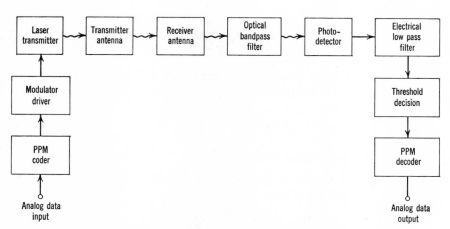

Figure 11-10   *PPM intensity modulation laser communication system*

† A more efficient decision rule, from the standpoint of carrier power utilization, is to assume that the carrier pulse has been transmitted in the time slot for which the electron count is largest. Implementation of the coder for this system is somewhat more difficult than with the threshold decoder.

output first equals or exceeds the threshold, or in a randomly selected time slot if the photodetector output does not equal or exceed the threshold.

There is a probability, $P_{SN}{}^P$, at each time slot that the threshold will be equalled or exceeded by the combination of the received signal and detector noise; there is also the probability, $P_N{}^P$, that the threshold will be equalled or exceeded by detector noise alone. The probability of making an incorrect detection decision may be expressed in terms of these signal and noise detection probabilities as

$$P_e{}^P = \left\{ \begin{array}{l} \text{probability that noise equals or exceeds threshold before a} \\ \text{signal slot} \end{array} \right\}$$

$$+ \left\{ \begin{array}{l} \text{probability that noise equals or exceeds threshold after} \\ \text{signal slot, given signal plus noise does not equal or} \\ \text{exceed threshold} \end{array} \right\}$$

$$+ \left\{ \begin{array}{l} \text{probability that neither noise nor signal plus noise} \\ \text{equals or exceeds threshold} \end{array} \right\}$$

$$\times \ \{\text{probability of an incorrect random choice}\}$$

Then,

$$P_e{}^P = \sum_{i=1}^{K} p_i [1 - (1 - P_N{}^P)^{i-1}] + \sum_{i=1}^{K} p_i (1 - P_N{}^P)^{i-1}(1 - P_{SN}{}^P)$$

$$\times \ [1 - (1 - P_N{}^P)^{K-i}] + (1 - P_N{}^P)^{K-1}(1 - P_{SN}{}^P) \sum_{i=1}^{K} p_i (1 - p_i) \quad (11\text{-}50)$$

where p_i is the *a priori* probability of transmitting a signal in the ith slot. For a uniform source distribution, i.e., $p_i = 1/K$, the probability of error is

$$P_e{}^P = \left(1 - \frac{P_{SN}{}^P}{K P_N{}^P} \right) + \frac{(1 - P_N{}^P)^{K-1}}{K P_N{}^P} (P_{SN}{}^P - P_N{}^P) \quad (11\text{-}51)$$

PPM/IM Shot Noise Limited Detection

For shot noise limited detection the detection probabilities, as given by Equations 9-20 and 9-21, are

$$P_{SN}{}^P = \sum_{k=k_D}^{\infty} \frac{[\mu_{S,P} + (\mu_{H,P}/K)]^k \exp\{-[\mu_{S,P} + (\mu_{H,P}/K)]\}}{k!} \quad (11\text{-}52)$$

and

$$P_N{}^P = \sum_{k=k_D}^{\infty} \frac{[(\mu_{H,P}/K)]^k \exp\{-[(\mu_{H,P}/K)]\}}{k!} \quad (11\text{-}53)$$

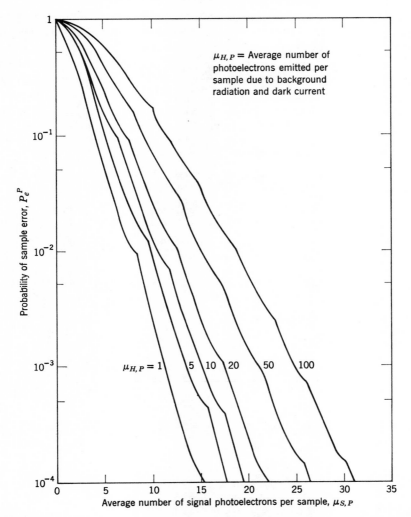

Figure 11-11   *Probability of detection error for PPM/IM direct detection system-shot noise limited operation*

where:

$\mu_{S,P}$ = average number of photoelectrons emitted per sample due to the laser signal

$\mu_{H,P}$ = average number of photoelectrons emitted per sample due to background radiation and dark current

The value of k_D that minimizes the probability of detection error is the greatest integer value of the likelihood ratio test threshold, k_T. The likelihood ratio test threshold, as given in Figure 9-7, is

$$k_T = \frac{\mu_{S,P} + \ln (K - 1)}{\ln [1 + (K\mu_{S,P}/\mu_{H,P})]} \tag{11-54}$$

Figure 11-11 gives the probability of detection error as a function of the signal and shot noise photoelectron counts per sampling period.

PPM/IM Thermal Noise Limited Detection

The detection probabilities for thermal noise limited detection, as noted in Equations 9-24 and 9-25, are

$$P_{SN}{}^P = \int_{i_F'}^{\infty} (2\pi\sigma_{i_T}{}^2)^{-1/2} \exp\left\{ -\frac{[i_F - (q/\tau_P)\mu_{S,P}]^2}{2\sigma_{i_T}{}^2} \right\} di_F \tag{11-55}$$

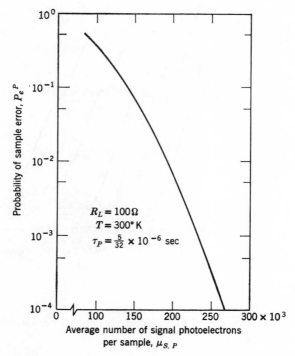

$$R_L = 100\,\Omega$$
$$T = 300°\,K$$
$$\tau_P = \frac{5}{32} \times 10^{-6} \text{ sec}$$

Probability of sample error, $P_e{}^P$

Average number of signal photoelectrons per sample, $\mu_{S,P}$

Figure 11-12   *Probability of detection error for PPM/IM direct detection laser communication system-thermal noise limited operation*

and

$$P_N{}^P = \int_{i_F{}'}^{\infty} (2\pi\sigma_{i_T}{}^2)^{-1/2} \exp\left\{-\frac{i_F{}^2}{2\sigma_{i_T}{}^2}\right\} di_F \qquad (11\text{-}56)$$

The detection threshold shown in Figure 9-9 is

$$i_F{}' = \frac{q}{2\tau_P} \mu_{S,P} + \frac{\sigma_{i_T}{}^2}{(q/\tau_P)\mu_{S,P}} \ln{(K-1)} \qquad (11\text{-}57)$$

Figure 11-12 gives the probability of detection error for the thermal noise limited PPM/IM communication system.

REFERENCES

11-1. Neblack, W. and Wolf, E. "Polarization Modulation and Demodulation of Light." *Applied Optics*, **3** (2), 277–279, Feb. 1964.

11-2. Pratt, W. K. and Norton, R. J. "An Experimental Optical Polarization Modulation Communication System." *1966 IEEE International Communications Conference*, 268–269, June 1966.

11-3. Pratt, W. K. "Binary Detection in an Optical Polarization Modulation Communication Channel." *IEEE Transactions on Communication Technology*, **COM-14** (5), 664–665, Oct. 1966.

11-4. Peters, W. N. and Arguello, R. J. "Fading and Polarization Noise of a PCM/PL System." *IEEE Journal of Quantum Electronics*, **QE-3** (11), 532–539, Nov. 1967.

11-5. Marcum, J. I. "A Statistical Theory of Target Detection by Pulsed Radar." *IRE Transactions on Information Theory*, **IT-6**, 159–160, Apr. 1960.

11-5. Schwartz, M., Bennett, W. R., and Stein, S. *Communication Systems and Techniques*. McGraw-Hill, New York, 1966.

11-7. "Tables of the *Q* Function." Rand Corp. Research Memo RM 399, Jan. 1950.

11-8. Brennan, L. E. and Reed, I. S. "A Recursive Method of Computing the *Q* Function." *IEEE Transactions on Information Theory*, **IT-11** (2), 312–313, Apr. 1965.

chapter 12

SUBCARRIER AND HETERODYNE DIGITAL LASER COMMUNICATION SYSTEMS

Expressions for the probability of detection error are derived in the following sections for a PCM/AM subcarrier intensity modulation system employing direct detection, a PCM intensity modulation heterodyne detection system, and a PCM frequency modulation heterodyne detection system. In the case of the subcarrier system, the noise at the subcarrier filter output is detector shot noise and thermal noise passed through the subcarrier filter. The latter noise is inherently Gaussian and the former approaches a Gaussian distribution for only moderately large emission rates due to the filtering action of the subcarrier filter. Hence the subcarrier filter output noise is modeled by a narrow-band Gaussian process. In the heterodyne detection systems the detector electron emission rate becomes quite high when the local oscillator amplitude is increased for shot noise limited detection. With such a high electron emission rate the IF filter output may be considered Gaussian.

12.1 PCM/AM SUBCARRIER INTENSITY MODULATION SYSTEM

Figure 12-1 contains a block diagram of a PCM/AM subcarrier intensity modulation system (PCM/AM/IM). The radio frequency subcarrier oscillator is set to its maximum amplitude to transmit a "one" bit and set to zero amplitude for a "zero" bit. The subcarrier oscillator wave modulates the laser intensity such that the laser is 100% intensity modulated by the subcarrier sine wave for a "one" bit and set at a constant half intensity level for a "zero" bit. The output of the electrical bandpass subcarrier filter is either filtered photodetector noise plus a sine wave of the subcarrier frequency or noise alone. Figure 12-2 illustrates waveforms in an experimental PCM/AM/IM system.

218

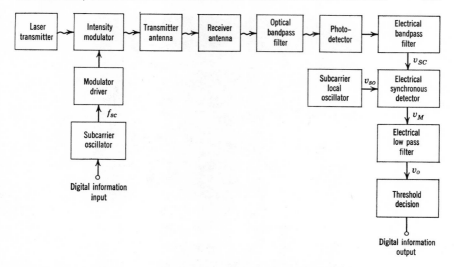

Figure 12-1 *PCM/AM subcarrier intensity modulation direct detection laser communication system*

From the standpoint of laser power utilization and signal shot noise reduction, it would be more efficient to inhibit completely the laser for a "zero." Strictly speaking such a system does not employ subcarrier modulation but rather direct carrier modulation with a signal set composed of no carrier and a carrier with a sine wave intensity variation. From a practical viewpoint the subcarrier modulation method initially described has the attributes that the carrier is always present at the receiver and its average power is constant. Under these circumstances the carrier may be used as a beacon to point the receiver toward the transmitter.

The instantaneous voltage, v_{SC}, at the subcarrier filter output, from Equation 10-17, is

$$v_{SC} = \frac{G\mathscr{D}P_C A_{SC} R_L}{2} \cos\left[\omega_{SC}t + \Phi_{SC}\right] \equiv A \cos\left[\omega_{SC}t + \Phi_{SC}\right] \quad (12\text{-}1)$$

where:

$$G = \text{detector current gain}$$

$$\mathscr{D} = \text{detector conversion factor } (\mathscr{D} = \eta q/hf_c)$$

$$P_C = \text{average carrier power}$$

$$A_{SC} = \text{subcarrier amplitude}$$

$$R_L = \text{optical receiver load resistance}$$

For a "one" bit, $A_{SC} = 1$, and for a "zero" bit, $A_{SC} = 0$.

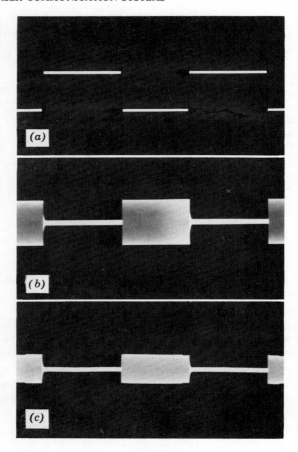

Figure 12-2 *PCM/AM subcarrier intensity modulation laser communication
system waveforms. (a) Digital information input, IK Hz square wave. (b) Modulator
driver input. (c) Electrical bandpass filter output*

Since the subcarrier radio frequency oscillator can be made quite stable
by crystal control, electrical synchronous detection is feasible using a phase-
locked loop controller for the radio frequency local oscillator. In the synchronous detector, when phase lock has been achieved, the subcarrier filter
output is multiplied by a sine wave

$$v_{SO} = A_{SO} \cos \left[\omega_{SC}t + \Phi_{SC} \right] \qquad (12\text{-}2)$$

The narrow-band Gaussian noise voltage input to the synchronous detector
may be written in quadrature form as

$$v_{n_{SC}} = X_c(t) \cos \left[\omega_{SC}t + \Phi_{SC} \right] - X_s(t) \sin \left[\omega_{SC}t + \Phi_{SC} \right] \qquad (12\text{-}3)$$

where the variances of the Gaussian random variables, X_c and X_s, are equal and related to the subcarrier noise power, $[N]_{SC}$, by

$$\sigma_X^2 = [N]_{SC} R_L \tag{12-4}$$

The product output, v_M, of the synchronous detector multiplier is

$$v_M \equiv v_{SO}(v_{SC} + v_{nsc}) \tag{12-5}$$

which, upon expansion, yields

$$v_M = \left[\frac{A A_{so}}{2} + \frac{X_c(t) A_{so}}{2}\right][1 + \cos(2\omega_{SC}t + 2\Phi_{SC})]$$

$$- \frac{X_s(t) A_{SO}}{2} \sin[2\omega_{SC}t + 2\Phi_{SC}] \tag{12-6}$$

A zonal low pass filter of bandwidth B_O following the multiplier removes the double frequency terms, leaving

$$v_O = \frac{A A_{SO}}{2} + \frac{X_c(t) A_{SO}}{2} \tag{12-7}$$

The quadrature noise component, $X_s(t)$, is thus eliminated by the synchronous detector. The subcarrier local oscillator amplitude, A_{SO}, multiplies the subcarrier signal amplitude, A, and the in-phase noise, $X_c(t)$, equally. Setting A_{SO} to a high amplitude swamps out any thermal noise due to resistive elements in the synchronous detector. Of course, thermal noise developed in the optical receiver still remains. The random variable

$$v_Z \equiv \frac{2}{A_{SO}} v_O = A + X_c(t) \tag{12-8}$$

has a Gaussian distribution with mean A since $X_c(t)$ is a zero mean Gaussian random variable. Hence, the probability densities at the low pass filter output for "one" and "zero" signals, as shown in Figure 12-3, are

$$P(v_Z \mid \bar{S}) = (2\pi[N]_{SC} R_L)^{-1/2} \exp\left\{-\frac{v_Z^2}{2[N]_{SC} R_L}\right\} \tag{12-9}$$

and

$$P(v_Z \mid S) = (2\pi[N]_{SC} R_L)^{-1/2} \exp\left\{-\frac{[v_Z - (G\mathscr{D} A_c^2 R_L/4)]^2}{2[N]_{SC} R_L}\right\} \tag{12-10}$$

For "ones" and "zeros" transmitted with equal probability, the optimum decision threshold found from a likelihood ratio test is

$$v_Z' = \frac{G\mathscr{D} A_c^2 R_L}{8} \tag{12-11}$$

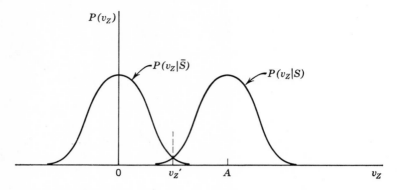

Figure 12-3   *Conditional probability densities for PCM/AM/IM direct detection system*

The probability of detection error is then

$$P_e{}^B = \tfrac{1}{2} \int_{-\infty}^{v_Z'} P(v_Z \mid S)\, dv_Z + \tfrac{1}{2} \int_{v_Z'}^{\infty} P(v_Z \mid \bar{S})\, dv_Z \qquad (12\text{-}12)$$

which can be written in terms of the Gaussian error function as

$$P_e{}^B = \frac{1}{2}\left[1 - \operatorname{erf}\left\{\frac{G\mathscr{D}P_C R_L}{4\sqrt{2[N]_{SC}R_L}}\right\}\right] \qquad (12\text{-}13)$$

For shot noise limited operation with $B_{SC} = 2B_O$, from Equation 10-19, the subcarrier noise power is

$$[N]_{SC} = 4qG^2\left[\mathscr{D}\frac{P_C}{2} + \mathscr{D}P_B + I_D\right]B_O R_L \qquad (12\text{-}14)$$

Setting the output filter bandwidth to be $B_O = 1/\tau_B$, where τ_B is the bit period, allows the probability of detection error to be written in terms of the signal and shot noise photoelectron counts per bit. The probability of detection error for shot noise limited detection, as shown in Figure 12-4, is then

$$P_e{}^B = \frac{1}{2}\left[1 - \operatorname{erf}\left\{\frac{1}{8}\frac{\mu_{S,B}}{\sqrt{\mu_{S,B} + 2\mu_{H,B}}}\right\}\right] \qquad (12\text{-}15)$$

where:

$\mu_{S,B}$ = average number of photoelectrons emitted per bit due to the laser signal

$\mu_{H,B}$ = average number of photoelectrons emitted per bit due to background radiation and dark current

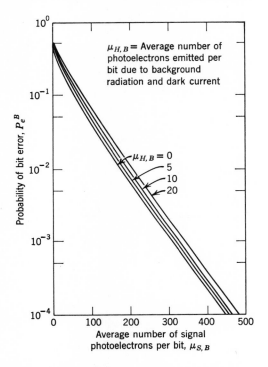

Figure 12-4 *Probability of detection error for PCM/AM/IM direct detection laser communication system-shot noise limited detection*

For ideal signal shot noise limited detection

$$P_e^B = \tfrac{1}{2}[1 - \mathrm{erf}\{\tfrac{1}{8}\sqrt{\mu_{S,B}}\}] \tag{12-16}$$

The subcarrier noise power for thermal noise limited detection, from Equation 10-20, is

$$[N]_{SC} = 8kTB_O \tag{12-17}$$

The probability of detection error for unity detector gain may then be written as

$$P_e^B = \frac{1}{2}\left[1 - \mathrm{erf}\left\{\frac{q\mu_{S,B}}{16}\sqrt{\frac{R_L}{kT\tau_B}}\right\}\right] \tag{12-18}$$

Figure 12-5 illustrates the probability of detection error for thermal noise limited detection.

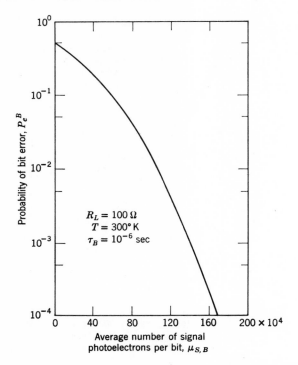

Figure 12-5   *Probability of detection error for PCM/AM/IM direct detection laser communication system-thermal noise limited detection*

12.2 PCM INTENSITY MODULATION HETERODYNE DETECTION SYSTEM

Figure 12-6 contains a block diagram of a heterodyne detection PCM intensity modulation system. The transmitter is the same as that for a PCM/IM direct detection system. At the output of the IF filter a sine wave of the IF frequency plus detector noise will be present if a "one" bit is transmitted; only detector noise will be present for the transmission of a "zero" bit. Two basic types of radio receivers can be employed. The most efficient one, called coherent AM, requires knowledge of the IF phase; the other type of radio receiver, incoherent AM, ignores the IF phase and simply performs threshold detection on the envelope of the IF signal.

The signal voltage at the output of the IF filter, v_{IF}, of a heterodyne detection optical receiver, as given by Equation 10-38, is

$$v_{\mathrm{IF}} = G\mathscr{D}A_c A_o R_L \cos\left[(\omega_o - \omega_c)t + (\Phi_o - \Phi_c)\right] \qquad (12\text{-}19)$$

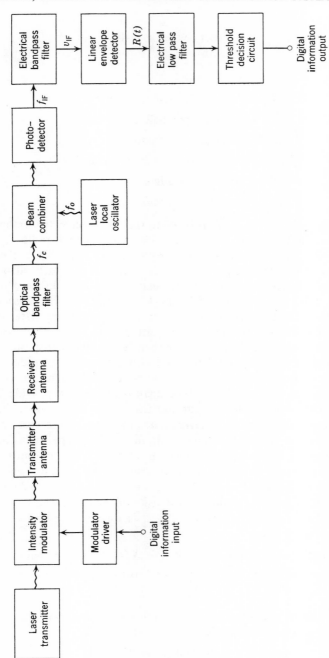

Figure 12-6 PCM/IM heterodyne detection laser communication system

or for simplicity

$$v_{\text{IF}} \equiv A \cos [\omega_{\text{IF}}t + \Phi_{\text{IF}}] \qquad (12\text{-}20)$$

where:

G = detector gain

$\mathscr{D}$ = detector conversion factor

A_c = carrier electric field amplitude

ω_c = carrier angular frequency

ω_o = local oscillator angular frequency

Φ_c = carrier phase angle

Φ_o = local oscillator phase angle

The IF phase angle, Φ_{IF}, is thus equal to the difference between the carrier and local oscillator phase angles. To perform coherent radio frequency detection, Φ_{IF} must be maintained constant or at least be tracked by a phase-locked loop. This requires that both the carrier and local oscillator phase angles be held nearly constant. Such operation is difficult to obtain because of stability limitations of lasers. If the necessary IF phase stability can be achieved, second detection can be performed by an electrical synchronous detection receiver. However, in this situation it would be more efficient to perform homodyne detection rather than heterodyne detection. For this reason only incoherent second detection techniques will be considered.

The incoherent AM receiver can be a linear envelope detector. A square-law device can also be employed since there is no need to provide linearity between the second detector output and the carrier amplitude or intensity. Linear envelope detection is preferred since the signal-noise cross products of the square-law detector are avoided. In the envelope detection radio receiver shown in Figure 12-6, the envelope of the IF filter output is formed by a linear envelope detector. A bit decision is based upon whether the amplitude of the envelope is greater or less than a decision threshold, which is preestablished from the probability density of the envelope.

The IF filter output noise voltage, $v_{n_{\text{IF}}}$, is assumed to be narrow-band Gaussian noise given by

$$v_{n_{\text{IF}}} = X_c(t) \cos [\omega_{\text{IF}}t + \Phi_{\text{IF}}] - X_s(t) \sin [\omega_{\text{IF}}t + \Phi_{\text{IF}}] \qquad (12\text{-}21)$$

where the variance of X_c and X_s is related to the IF noise power, $[N]_{\text{IF}}$, by

$$\sigma_X{}^2 = [N]_{\text{IF}}R_L \qquad (12\text{-}22)$$

The sum of the IF signal voltage, v_{IF}, and the IF noise voltage, $v_{n_{\text{IF}}}$, can then be expressed in phase-magnitude form as

$$v_{\text{IF}} + v_{n_{\text{IF}}} = R(t) \cos [\omega_{\text{IF}}t + \Phi_{\text{IF}} + \psi(t)] \qquad (12\text{-}23)$$

where the envelope of the IF signal and noise is

$$R(t) = \sqrt{[X_C(t) + A]^2 + X_s^2(t)} \qquad (12\text{-}24)$$

The phase angle $\psi(t)$ can also be expressed in terms of the signal and noise parameters, but it is not needed for the following analysis because the envelope detector output is $R(t)$ only. From Equation B-52 the conditional probability densities of the envelope for a carrier transmitted and no carrier, as shown in Figure 12-7, are

$$P(R \mid S) = \frac{R}{\sigma_X^2} \exp\left\{-\frac{R^2 + A^2}{2\sigma_X^2}\right\} I_0\left\{\frac{AR}{\sigma_X^2}\right\} \qquad (12\text{-}25)$$

and

$$P(R \mid \bar{S}) = \frac{R}{\sigma_X^2} \exp\left\{-\frac{R^2}{2\sigma_X^2}\right\} \qquad (12\text{-}26)$$

The likelihood ratio test yields the decision rule that the IF signal is to be judged present if

$$\frac{P(R \mid S)}{P(R \mid \bar{S})} \geq \frac{1 - p}{p} \qquad (12\text{-}27)$$

where p is the *a priori* probability of transmitting the carrier. For "ones" and "zeros" transmitted with equal probability the likelihood ratio test envelope threshold, R_T, is found from the solution of the transcendental equation

$$\exp\left\{-\frac{A^2}{2\sigma_X^2}\right\} = I_0\left\{\frac{AR}{\sigma_X^2}\right\} \qquad (12\text{-}28)$$

A good approximation to the solution of this equation is [12-1]

$$R_T \approx \sigma_X\sqrt{2 + \frac{A^2}{4\sigma_X^2}} \qquad (12\text{-}29)$$

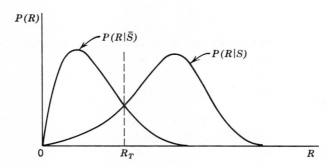

Figure 12-7 *Conditional probability densities for PCM/FM heterodyne detection system*

The probability that the envelope of the signal plus noise exceeds the detector threshold and the probability that the envelope of the noise alone exceeds the detector threshold are

$$P_{SN}{}^B = \int_{R_T}^{\infty} \frac{R}{\sigma_X{}^2} \exp\left\{ -\frac{(R^2 + A^2)}{2\sigma_X{}^2} \right\} I_0\left(\frac{AR}{\sigma_X{}^2}\right) dR \qquad (12\text{-}30)$$

and

$$P_N{}^B = \int_{R_T}^{\infty} \frac{R}{\sigma_X{}^2} \exp\left\{ -\frac{R^2}{2\sigma_X{}^2} \right\} dR \qquad (12\text{-}31)$$

The first integral can be expressed in terms of Marcum's Q functions [12-1]

$$P_{SN}{}^B = Q\left[\frac{A}{\sigma_X}, \frac{R_T}{\sigma_X}\right] \qquad (12\text{-}32)$$

where

$$Q(a, b) \equiv \int_{b}^{\infty} \exp\left\{ -\frac{(a^2 + x^2)}{2} \right\} I_0(ax)x\, dx \qquad (12\text{-}33)$$

The second integral may be evaluated directly or written in terms of the Q function

$$P_N{}^B = \exp\left\{ -\frac{R_T{}^2}{2\sigma_X{}^2} \right\} = Q\left[0, \frac{R_T}{\sigma_X}\right] \qquad (12\text{-}34)$$

The probability of detection error for "ones" and "zeros" transmitted with equal likelihood is then from Equation 11-2, equal to

$$P_e{}^B = \frac{1}{2}\left\{ 1 + Q\left[0, \frac{R_T}{\sigma_X}\right] - Q\left[\frac{A}{\sigma_X}, \frac{R_T}{\sigma_X}\right] \right\} \qquad (12\text{-}35)$$

The ratio $A/2\sigma_X{}^2$ is equal to the IF signal-to-noise ratio which, for a high level of local oscillator power (from Equation 10-43), is

$$\left[\frac{S}{N}\right]_{\mathrm{IF}} = \frac{\eta P_C}{hf_c B_{\mathrm{IF}}} \qquad (12\text{-}36)$$

For $B_{\mathrm{IF}} = 2B_O = 2/\tau_B$, using the approximation of Equation 12-29, the probability of detection error becomes

$$P_e{}^B = \frac{1}{2}\left\{ 1 + Q\left[0, \sqrt{2 + \frac{\mu_{S,B}}{4}}\right] - Q\left[\sqrt{\mu_{S,B}}, \sqrt{2 + \frac{\mu_{S,B}}{4}}\right] \right\} \qquad (12\text{-}37)$$

Figure 12-8 is a plot of the probability of detection error for the PCM/AM heterodyne detection system. For high signal electron counts the probability of error expression becomes

$$P_e{}^B \approx \frac{1}{2} \exp\left\{ \frac{\mu_{S,B}}{8} \right\} \qquad (12\text{-}38)$$

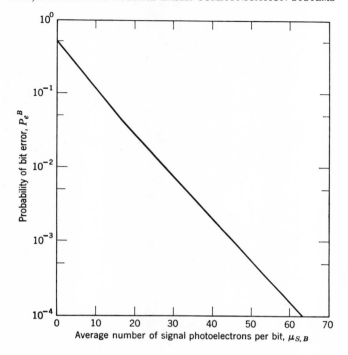

Figure 12-8   *Probability of detection error for PCM/IM heterodyne detection laser communication system*

12.3 PCM FREQUENCY MODULATION HETERODYNE DETECTION SYSTEM

In the optical PCM frequency modulation system, information is conveyed by transmission of the laser carrier at one of two different frequencies to represent "one" and "zero" bits. A block diagram of a typical PCM/FM laser communication system employing heterodyne detection is shown in Figure 12-9. The output of a continuous wave laser is shifted to one of two transmission frequencies by a frequency modulator under control of the data. At the receiver the incident signal is spatially mixed with a local oscillator laser in an optical heterodyne receiver to shift the information to the inter-mediate frequency. An incoherent twin filter radio FM receiver then demodu-lates the IF frequency signal. The radio receiver consists of a pair of electrical bandpass filters tuned to the carrier frequencies associated with the "one" and "zero" bits. A linear envelope detector following each bandpass filter provides an output signal proportional to the envelope of the filter output.

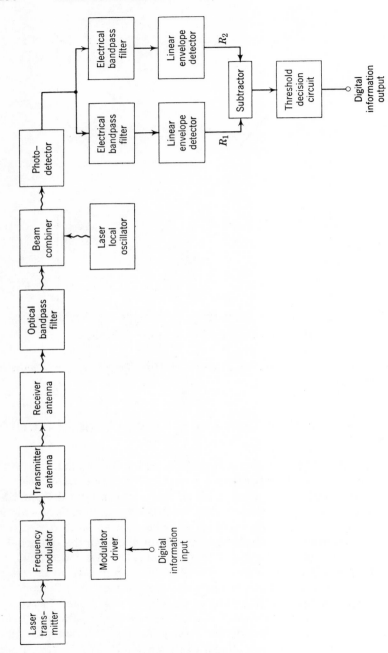

Figure 12-9 *PCM/FM heterodyne detection laser communication system*

The envelope detector outputs are subtracted and a bit decision is made on the amplitude of the difference signal.

The signal voltage at the photodetector output for a "one" bit is

$$v_{IF} = A \cos [(\omega_0 - \omega_c - \omega_d)t + (\Phi_o - \Phi_c)] \tag{12-39}$$

and for a "zero" bit is

$$v_{IF} = A \cos [(\omega_0 - \omega_c + \omega_d)t + (\Phi_o - \Phi_c)] \tag{12-40}$$

where $A \equiv G\mathscr{D}A_cA_oR_L$ and ω_d is the angular frequency deviation.

The output of one of the bandpass filters is narrow-band Gaussian noise plus the IF signal; the output of the other is narrow-band Gaussian noise only. Hence the conditional probability densities at the output of the envelope detectors may be written as

$$P(R_1 \mid S_1) = P(R_2 \mid S_2) = \frac{R_i}{\sigma_X{}^2} \exp \left\{ -\frac{(R_i{}^2 + A^2)}{2\sigma_X{}^2} \right\} I_0 \left\{ \frac{AR_i}{\sigma_X{}^2} \right\} \tag{12-41}$$

and

$$P(R_2 \mid S_1) = P(R_1 \mid S_2) = \frac{R_i}{\sigma_X{}^2} \exp \left\{ -\frac{R_i{}^2}{2\sigma_X{}^2} \right\} \tag{12-42}$$

where $\sigma_X{}^2 = [N]_{IF}R_L$ is the noise variance at the IF filter output. For ones and zeros transmitted with equal probability, by symmetry of the channels, the likelihood ratio test threshold of the difference signal $R_2 - R_1$ is zero. Hence, an error occurs if $R_2 > R_1$ if S_1 is transmitted, and if $R_1 > R_2$ if S_2 is transmitted. Since the noise is uncorrelated between channels, the joint density of R_1 and R_2 is

$$P(R_1, R_2 \mid S_i) = P(R_1 \mid S_i)P(R_2 \mid S_i) \tag{12-43}$$

The probability of detection error then becomes

$$P_e{}^B = P(R_2 > R_1) = \int_{R_1 = 0}^{\infty} \int_{R_2 = R_1}^{\infty} \frac{R_1}{\sigma_X{}^2} \exp \left\{ -\frac{(R_1{}^2 + A^2)}{2\sigma_X{}^2} \right\} I_0 \left\{ \frac{AR_1}{\sigma_X{}^2} \right\}$$
$$\times \frac{R_2}{\sigma_X{}^2} \exp \left\{ -\frac{R_2{}^2}{2\sigma_X{}^2} \right\} dR_1 \, dR_2 \tag{12-44}$$

Integrating with respect to R_1 yields

$$P_e{}^B = \int_{R_1 = 0}^{\infty} \frac{R_1}{\sigma_X{}^2} \exp \left\{ -\frac{(2R_1{}^2 + A^2)}{2\sigma_X{}^2} \right\} I_0 \left(\frac{AR_1}{\sigma_X{}^2} \right) dR_1 \tag{12-45}$$

which reduces to the surprisingly simple form

$$P_e{}^B = \frac{1}{2} \exp \left\{ -\frac{A^2}{4\sigma_X{}^2} \right\} \tag{12-46}$$

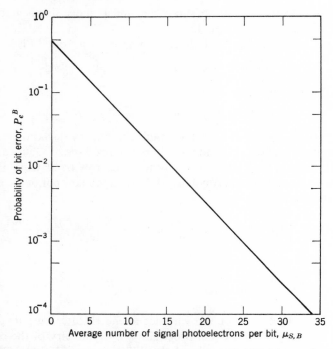

Figure 12-10 *Probability of detection error for PCM/FM heterodyne detection laser communication system*

The exponential argument is the IF signal-to-noise ratio. For a strong local oscillator the probability of error can be written as

$$P_e^B = \frac{1}{2} \exp\left\{ -\frac{\mu_{S,B}}{4} \right\} \qquad (12\text{-}47)$$

Figure 12-10 is a plot of the probability of detection error for the PCM/FM system.

REFERENCE

12-1. Schwartz, M., Bennett, W. R., and Stein, S. *Communication Systems and Techniques.* McGraw-Hill, New York, 1966.

chapter 13

OPTIMUM SYSTEM DESIGN

The complexity of laser communication systems demands that a unified methodical approach be taken to their optimum design. Beyond the obvious question as to what modulation or detection method to use, the designer must pick an operating frequency and designate the system parameters such as laser transmitter power, transmitter aperture diameter, receiver aperture diameter, and receiver field of view. These choices must ultimately be made on the basis of minimizing system cost.

13.1 OPTIMIZATION METHODOLOGY

For different types of communication systems and different missions, the basis for optimization may vary. But, in general, the system designer seeks the greatest information fidelity, maximum communication range, and maximum information rate for the least total system cost. System cost includes the fabrication cost of the system components plus, in the case of space missions, the expense of placing the components aboard a spacecraft on a cost per unit weight basis. With analog systems, information fidelity is generally related to signal-to-noise ratio. Pulse and digital systems are usually rated in terms of the probability of detection error per sample or bit. In this chapter optimization techniques are developed only for pulse and digital systems. The results are easily extended for analog systems.

The information required for an optimum system design, as shown in Figure 13-1, is derived from: (1) an analysis of the fabrication cost, weight, and power supply requirements of the communication components, and (2) a communication system analysis which provides the relationship between the communication parameters, noise effects, and the operating environment.

The optimization procedure is to develop relationships between the cost of the system components and the values of the major system parameters-transmitter power, transmitter antenna diameter, receiver antenna diameter, and receiver field of view. Then for the particular receiver employed and for the specified operational environment, the expression for the receiver probability of detection error is minimized, consistent with the cost relationships,

233

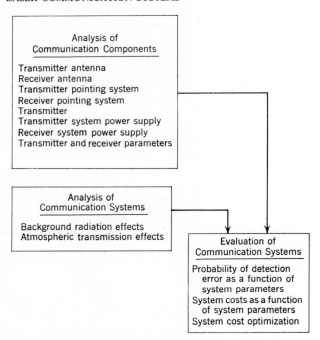

Figure 13-1   Systems optimization flow chart

under the constraint that the total system cost remains fixed. Since the communication range and information rate are monotonic functions of the probability of detection error, they are maximized when the probability of detection error is minimized.

The optimization procedure described in this chapter is admittedly complex to perform though not complicated in principle. As with any optimization problem, the accuracy of the solution is dependent upon the accuracy of the cost functions. It may be argued that the estimation of component cost functions requires too much guesswork and, therefore, that the solution of the optimization problem is suspect. The counter argument is that the alternative to the optimization procedure is the selection of the system parameters completely by guesswork or trial and error. From a practical standpoint it has been found that reasonably accurate component cost functions can be modeled; moreover the optimization problem solutions are not overly sensitive to the accuracy of the cost functions [13-1 to 13-3].

Analysis of Communication Components

The fabrication cost, weight, and power requirement burden relations are expressed only as functions of the major system parameters—transmitter antenna diameter, transmitter power, receiver antenna diameter, and receiver field of view. Transmission wavelength, information rate, and other operating conditions, such as whether the transmitter or receiver is ground based or aboard a spacecraft, determine the nature of the functional relationships between the major system parameters and component burdens. In actual practice the relations are modeled by power series for ease in manipulation and computation. The functional relations between the burdens and system parameters are listed in Table 13-1 and described below.

Component	Burden	Parameter Relationship			
		d_T	P_L	d_R	θ_R
Transmitter antenna	Weight	×			
	Fabrication cost	×			
Receiver antenna	Weight			×	
	Fabrication cost			×	
Transmitter antenna pointing system	Weight	×			
	Fabrication cost	×			
	Power requirement	×			
Receiver antenna pointing system	Weight			×	
	Fabrication cost				×
	Power requirement			×	
Transmitter	Weight		×		
	Fabrication cost		×		
	Power requirement		×		
Transmitter system power supply	Weight	×	×		
	Fabrication cost				
Receiver system power supply	Weight			×	
	Fabrication cost			×	

Table 13-1. OPTICAL COMMUNICATION SYSTEM BURDEN RELATIONSHIPS

Transmitter Antenna. The weight and fabrication cost of a transmitter optical antenna system are dependent upon the transmitter antenna diameter. A transmitter antenna is usually designed to operate as close to the diffraction limit as possible to achieve the greatest spatial power density at the receiver for a given transmitter antenna diameter. For small transmitter antennas, the weight is proportional to the antenna area and hence to the square of the antenna diameter. For larger antennas, as structural supports are added to maintain the rigidity required for diffraction limited operation, the weight dependence becomes volumetric.

Receiver Antenna. The weight and fabrication cost of a receiver optical antenna system are dependent upon the receiver antenna diameter. Since a receiver antenna is not normally designed to be diffraction limited, its construction and mechanical support tolerances need not be as stringent as for a transmitter antenna.

Transmitter Antenna Pointing System. A typical transmitter antenna pointing system consists of a gimballed support unit, which holds the transmitter antenna, and an associated control system, which points the transmitter antenna toward the receiver. The weight of the transmitter antenna pointing system is relatively insensitive to the transmitter pointing accuracy. Its weight is proportional to the weight of the transmitter antenna whose weight is dependent upon the transmitter antenna diameter.

The fabrication cost of the transmitter pointing equipment is inversely proportional to the transmitter pointing accuracy. Pointing accuracy is usually specified as a fixed percentage of the transmitter beamwidth. Since the transmitter antenna is diffraction limited, the fabrication cost is proportional to the transmitter aperture diameter. The electrical power requirement for the transmitter antenna pointing system is primarily dependent upon the weight of the transmitter antenna.

Receiver Antenna Pointing System. The weight of the receiver pointing system is relatively insensitive to the receiver pointing accuracy. Its weight is proportional to the weight of the receiver antenna, which is itself dependent upon the receiver antenna diameter. The fabrication cost of the receiver pointing equipment is inversely proportional to the receiver pointing accuracy which is a fixed percentage of the receiver field of view. The power supply requirement for the receiver pointing system is primarily dependent upon the weight of the receiver antenna.

Transmitter. For a given transmission wavelength, within limits, the weight and fabrication cost of a laser transmitter are dependent upon the transmitter power. The electrical input power requirement is directly proportional to the transmitted power.

Transmitter System Power Supply. The fabrication cost and weight of the electrical power supply and conversion equipment at the transmitter are dependent upon the electrical power requirements of the transmitter antenna pointing system, transmitter, and modulator.

Receiver System Power Supply. The fabrication cost and weight of the electrical power supply and conversion equipment at the receiver are dependent upon the power requirements of the receiver pointing system and communication receiver equipment.

Transmitter and Receiver Parameters. The transmitter and receiver are characterized by the following parameters:

η = photodetector quantum efficiency
G = photodetector current gain
I_D = dark current
R_L = receiver load resistance
B_i = optical input filter bandwidth
B_O = receiver output filter bandwidth
τ_t = transmitter transmissivity
τ_r = receiver transmissivity
T = receiver temperature

Analysis of Communication Systems

Background Radiation Effects. Background radiation is expressed as a power spectral density in both frequency and space. The background radiation power input to the optical receiver is found by integrating the background radiation spectral irradiance over the input filter bandwidth and the receiver field of view.

Atmospheric Transmission Effects. Optical signals traveling through the atmosphere experience a transmission loss due to absorption and scattering by particles in the atmosphere. This loss is described by the atmospheric transmissivity, τ_a, whose value is dependent upon the transmission wavelength. In the case of heterodyne or homodyne optical detection, atmospheric turbulence places a limit on the size of the receiving aperture.

Evaluation of Communication Systems

Probability of Detection Error. For each type of receiver the probability of detection error may be expressed as a function of the laser transmitter power, transmitter antenna diameter, receiver antenna diameter, receiver field of view, receiver parameters, background radiation, receiver temperature, receiver bandwidth, transmission path, transmission wavelength, and communication range.

System Cost as a Function of System Parameters. The total system cost of the communication system, dependent upon the major system parameters, is the sum of the individual component fabrication costs plus the weight cost of placing each component aboard a spacecraft for a spaceborne system. This total system cost often may be subdivided to yield composite functional relations dependent upon each of the major system parameters. Let:

C_T = that portion of the total system cost, dependent upon the transmitter antenna diameter only, which is composed of the fabrication and weight cost of the transmitter antenna and the fabrication and weight cost of the transmitter pointing equipment and its associated power supply

C_R = that portion of the total system cost, dependent upon the receiver antenna diameter only, which is composed of the fabrication and weight cost of the receiver antenna and the weight cost of receiver pointing equipment and its associated power supply

C_Q = that portion of the total system cost, dependent upon receiver field of view only, which is composed of the fabrication cost of the receiver pointing equipment

C_G = that portion of the total system cost, dependent upon the transmitter power only, which is composed of the fabrication and weight cost of the transmitter and its associated power supply.

The sum of the system costs is defined as the optimization cost, C_V.

$$C_V \equiv C_T + C_R + C_G + C_Q \tag{13-1}$$

System Cost Optimization. The system cost allotments may be optimized by minimizing the probability of detection error as a function of the major system parameters for a fixed communication range under the constraint that the optimization cost, C_V, remains constant. By the method of Lagrange multipliers [13-4] the dummy expression

$$Q = P_e + \Lambda(C_V - C_T - C_R - C_G - C_Q) \tag{13-2}$$

if formed where Q and Λ are dummy variables. Then the partial derivatives of Q with respect to each of the system parameters (transmitter antenna diameter, d_T; receiver antenna diameter, d_R; transmitter power, P_L; and receiver field of view, θ_R) are set equal to zero.

$$\frac{\partial Q}{\partial d_T} = \frac{\partial P_e}{\partial d_T} - \Lambda \frac{\partial C_T}{\partial d_T} = 0 \tag{13-3}$$

$$\frac{\partial Q}{\partial d_R} = \frac{\partial P_e}{\partial d_R} - \Lambda \frac{\partial C_R}{\partial d_R} = 0 \tag{13-4}$$

$$\frac{\partial Q}{\partial P_L} = \frac{\partial P_e}{\partial P_L} - \Lambda \frac{\partial C_G}{\partial P_L} = 0 \tag{13-5}$$

$$\frac{\partial Q}{\alpha \theta_R} = \frac{P_e}{\theta_R} - \Lambda \frac{\partial C_Q}{\partial \theta_R} = 0 \tag{13-6}$$

Equating the Λ's yields

$$\frac{\partial P_e / \partial d_T}{\partial C_T / \partial d_T} = \frac{\partial P_e / \partial P_L}{\partial C_G / \partial P_L} = \frac{\partial P_e / \partial \theta_R}{\partial C_Q / \partial \theta_R} = \frac{\partial P_e / \partial d_R}{\partial C_R / \partial d_R} \tag{13-7}$$

The simultaneous solutions of these equations yields expressions for the optimum values of the system parameters in terms of the optimization cost, C_V. For a given value of C_V the optimum system parameters give the highest possible information rate and communication range for an arbitrary value of the signal-to-noise ratio. The solution of Equation 13-7 is easily found by recursive techniques on a digital computer. Solutions are possible when the system cost functions, C_T, C_R, C_G, and C_Q, are known only graphically. Thus the difficulty of mathematically modeling the system cost functions may be avoided. If one system parameter is held constant (e.g., a fixed receiver antenna diameter) or if one parameter is bounded in size by technological limitations, the characteristic equation or equations containing the constant parameter are simply eliminated from the simultaneous solution.

13.2 OPTIMUM SYSTEM DESIGN EXAMPLE

As an example of the optimization methodology procedure, the optimum design is determined for a thermal noise limited, direct detection, optical, digital communication system operating at a transmission wavelength of 10.6 microns transmitting data from a spacecraft to an earth receiver at a range of 10^6 km. The optimization is based upon the minimization of the transmitter system weight for a fixed parameter receiver.

Communication Components Analysis. Figures 13-2 to 13-5 give the weight and power requirements of the transmitter system components as a function of the transmitter antenna diameter and transmitter power required. At a transmission wavelength of 10.6 microns, the pertinent system parameters are:

$$\begin{array}{ll}
\text{photodetector quantum efficiency } \eta & = 0.50 \\
\text{transmitter transmissivity, } \tau_t & = 0.80 \\
\text{receiver transmissivity, } \tau_r & = 0.70 \\
\text{receiver temperature, } T & = 300^\circ \text{ K} \\
\text{receiver load resistance } R_L & = 100 \ \Omega
\end{array}$$

Communication Systems Analysis. Background radiation shot noise is not of concern since the system is thermal noise limited. At a transmission wavelength of 10.6 microns the atmospheric transmissivity is $\tau_a = 0.70$.

Communication Systems Evaluation. The probability of detection error for a thermal noise limited direct detection optical receiver, as given by Equation 11-13, is

$$P_e = \frac{1}{2}\left\{1 - \text{erf}\left(\frac{q\mu_{s,B}}{4}\sqrt{\frac{R_L}{2kT\tau_B}}\right)\right\} \tag{13-8}$$

where $\mu_{s,B}$ is the average number of signal electrons emitted by the detector per bit and τ_B is the bit period.

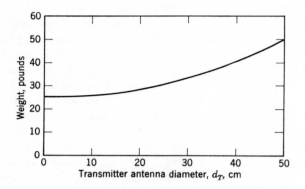

Figure 13-2   Transmitter antenna weight

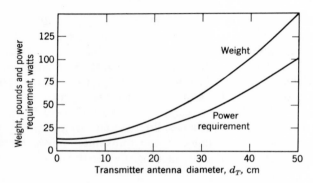

Figure 13-3   Transmitter pointing equipment weight and power requirement

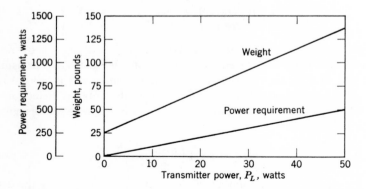

Figure 13-4   Transmitter weight and power requirement

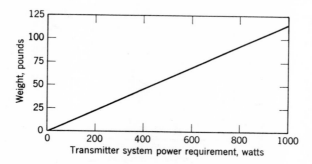

Figure 13-5   *Transmitter system power supply weight*

The signal photoelectron count is related to the average carrier power, P_C, by

$$\mu_{S,B} = \frac{\eta \tau_B}{h f_c} P_C \qquad (13\text{-}9)$$

Then, using the expression for the carrier power given by Equation 1-16, the probability of error may be written as

$$P_e = \tfrac{1}{2}\{1 - \operatorname{erf}(K_T d_T{}^2 d_R{}^2 P_L)\} \qquad (13\text{-}10)$$

where

$$K_T \equiv \frac{\eta q}{128 h f_c} \sqrt{\frac{R_L \tau_B}{2kT}} \frac{\tau_t \tau_a \tau_r}{\lambda^2 R^2}$$

Figures 13-6 and 13-7 give the system cost functions for a fixed weight

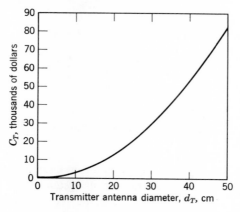

Figure 13-6   *Transmitter antenna system cost, C_T*

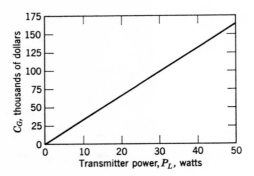

Figure 13-7 *Transmitter system cost, C_G*

cost of $1000 per pound to place the transmitter components aboard a spacecraft.

The optimization is performed by the method of Lagrange multipliers. The partial derivatives of the probability of error expression with respect to the major system parameters are

$$\frac{\partial P_e}{\partial d_T} = -2K_T d_T d_R{}^2 P_L \exp\{-K_T d_T{}^2 d_R{}^2 P_L\} \qquad (13\text{-}11)$$

$$\frac{\partial P_e}{\partial P_L} = -K_T d_T{}^2 d_R{}^2 \exp\{-K_T d_T{}^2 d_R{}^2 P_L\} \qquad (13\text{-}12)$$

From these equations the following set of characteristic equations is obtained

$$d_T \frac{\partial C_T}{\partial d_T} - 2P_L \frac{\partial C_G}{\partial P_L} = 0 \qquad (13\text{-}13)$$

$$C_T + C_R + C_G + C_Q = C_V \qquad (13\text{-}14)$$

A method of solution for the characteristic equations for a fixed optimization cost is to:

1. Select a trial value of d_T.
2. Determine P_L from Equation 13-13.
3. If Equation 13-14 is satisfied, stop; otherwise repeat step 1 for a new trial value of d_T.

Figure 13-8 gives the value of the optimum system parameters as a function of information rate for the communication system when the receiver field of view is set at $\theta_R = 10^{-3}$ radians and the receiver antenna diameter is set at $d_R = 5$ meters. The transmitter power is shown unbounded in one

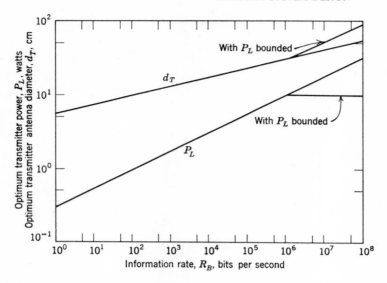

Figure 13-8   *Optimum system parameters for transmitter system, weight optimized, thermal noise limited, direct detection laser communication system with 500 cm receiver antenna diameter*

case and bounded at a maximum value of 10 watts in the other. The probability of detection error is set at 10^{-4} in the example.

REFERENCES

13-1. "Parametric Analysis of Microwave and Laser Systems for Communication and Tracking," Hughes Aircraft Co., Report No. P67-09, Dec. 1966.

13-2. Graves, R. E. "Techniques for Planning R and D for Deep Space Communications." *Proceedings AIAA 4th Annual Meeting*, Anaheim, Cal., Paper No. 67-973, Oct. 1967.

13-3. Breese, M. E. and Sferrazza, P. J. "Optimization of System Parameters for Deep Space Communication Systems." *IRE Globecom Convention Record*, pp. 180–183, 1961.

13-4. Schechter, R. S. *The Variational Method in Engineering*. McGraw-Hill, New York, 1967.

appendix A

POLARIZATION MATRIX ALGEBRA

The effect of optical components such as polarizers, wave plates, and attenuators on plane wave monochromatic light is conveniently described by the Jones calculus [A-1]. With the use of a complex representation of the electric field, the polarization state of light is denoted by a column vector

$$\mathscr{L} = \begin{bmatrix} E_X \\ E_Y \end{bmatrix} \tag{A-1}$$

where E_X and E_Y are the orthogonal components of the electrical field. Table A-1 gives polarization matrices for various polarization states.

Polarization State	Polarization Matrix $\mathscr{L}$	Coherency Matrix J
Linear polarization along horizontal axis	$\begin{bmatrix} 1 \\ 0 \end{bmatrix}$	$\begin{bmatrix} 1 & 0 \\ 0 & 0 \end{bmatrix}$
Linear polarization along vertical axis	$\begin{bmatrix} 0 \\ 1 \end{bmatrix}$	$\begin{bmatrix} 0 & 0 \\ 0 & 1 \end{bmatrix}$
Linear polarization at 45° to horizontal axis	$\dfrac{1}{\sqrt{2}} \begin{bmatrix} 1 \\ 1 \end{bmatrix}$	$\dfrac{1}{2} \begin{bmatrix} 1 & 1 \\ 1 & 1 \end{bmatrix}$
Linear polarization at $-45°$ to horizontal axis	$\dfrac{1}{\sqrt{2}} \begin{bmatrix} 1 \\ -1 \end{bmatrix}$	$\dfrac{1}{2} \begin{bmatrix} 1 & -1 \\ -1 & 1 \end{bmatrix}$
Right circular polarization	$\dfrac{1}{\sqrt{2}} \begin{bmatrix} -i \\ 1 \end{bmatrix}$	$\dfrac{1}{2} \begin{bmatrix} 1 & -i \\ i & 1 \end{bmatrix}$
Left circular polarization	$\dfrac{1}{\sqrt{2}} \begin{bmatrix} i \\ 1 \end{bmatrix}$	$\dfrac{1}{2} \begin{bmatrix} 1 & i \\ -i & 1 \end{bmatrix}$

Table A-1. POLARIZATION MATRICES OF POLARIZED LIGHT

In the polarization matrix algebra, passive optical components are expressed as operators which perform linear operations on the light input vectors. Optical system components are represented as

$$M = \begin{bmatrix} m_{11} & m_{12} \\ m_{21} & m_{22} \end{bmatrix} \tag{A-2}$$

where the matrix element m_{ij} describes the attenuation and phase shifting properties of the component. Table A-2 lists polarization matrices of some optical components.

Component	Polarization Matrix
Phase shifter (phase shift of 2Γ between vectors)	$\begin{bmatrix} e^{i\Gamma} & 0 \\ 0 & e^{-i\Gamma} \end{bmatrix}$
Polarization rotator (rotation of plane of polarization by angle θ)	$\begin{bmatrix} \cos\theta & -\sin\theta \\ \sin\theta & \cos\theta \end{bmatrix}$
Polarizer (projection of E field at angle θ to horizontal axis)	$\begin{bmatrix} \cos^2\theta & \sin\theta\cos\theta \\ \sin\theta\cos\theta & \sin^2\theta \end{bmatrix}$
Attenuator (absorption coefficients of ϵ_X and ϵ_Y along horizontal and vertical axes)	$\begin{bmatrix} e^{-\epsilon_X} & 0 \\ 0 & e^{-\epsilon_Y} \end{bmatrix}$

Table A-2. POLARIZATION MATRICES OF OPTICAL COMPONENTS

The polarization matrix of the light output is

$$\mathscr{L}' = M\mathscr{L} = \begin{bmatrix} E_X' \\ E_X' \end{bmatrix} \tag{A-3}$$

where:

$$E_X' = m_{11}E_X + m_{12}E_Y$$

$$E_Y' = m_{21}E_X + m_{22}E_Y$$

The Jones algebra is convenient for representing completely polarized light (i.e., linear or circular polarization). For light that is unpolarized or partially polarized, a coherency matrix algebra is employed. The coherency matrix of a plane wave is [A-2]

$$J = \begin{bmatrix} \overline{E_X E_X^*} & \overline{E_X E_Y^*} \\ \overline{E_Y E_X^*} & \overline{E_Y E_Y^*} \end{bmatrix} \tag{A-4}$$

where E_X^* and E_Y^* represent the complex conjugates of the electric field vectors. Each entry of the coherency matrix describes the time correlation of the field components through the time average of the electric field vectors. Table A-1 lists the coherency matrix for various polarization states. The coherency matrix of light passing through an optical component described by a polarization matrix, M, is

$$J = MJM^\dagger \tag{A-5}$$

where $M^\dagger$ is the Hermitian conjugate of M (i.e., complex conjugate of each entry of matrix transpose). The trace of the coherency matrix is the field intensity.

$$\mathscr{I} = T_r[J] = \widetilde{E_X E_X^*} + \widetilde{E_Y E_Y^*} \tag{A-6}$$

REFERENCES

A-1. Shurcliff, W. A. and Ballard, S. S. *Polarized Light*. Van Nostrand Momentum Books No. 7, Princeton, N.J., 1964.

A-2. O'Neill, E. L. *Introduction to Statistical Optics*. Addison-Wesley, Reading, Mass., 1963.

appendix B

GAUSSIAN PROCESS

B.1. GAUSSIAN RANDOM VARIABLE

A discrete random variable X is Gaussian distributed with parameters μ_X and σ_X if its probability density function, $f(X)$, is of the form

$$f(X) = \frac{1}{\sqrt{2\pi\sigma_X^2}} \exp\left\{-\frac{(X - \mu_X)^2}{2\sigma_X^2}\right\} \tag{B-1}$$

The corresponding cumulative probability distribution function, $F(X)$, is

$$F(X) = \int_{-\infty}^{X} \frac{1}{\sqrt{2\pi\sigma_X^2}} \exp\left\{-\frac{(Y - \mu_X)^2}{2\sigma_X^2}\right\} dY \tag{B-2}$$

Figure B-1 contains a sketch of the Gaussian density and distribution functions. The Gaussian error function, erf $\{X\}$, is defined as

$$\text{erf}\{X\} = \frac{2}{\sqrt{\pi}} \int_0^X \exp\{-Y^2\} dY \tag{B-3}$$

The error function has the following properties:

$$\text{erf}\{-X\} = -\text{erf}\{X\} \tag{B-4}$$

$$\text{erf}\{0\} = 0 \tag{B-5}$$

$$\text{erf}\{\infty\} = 1 \tag{B-6}$$

The following are useful identities involving the Gaussian error function.

$$\int_{-\infty}^{X} \frac{1}{\sqrt{2\pi\sigma_X^2}} \exp\left\{-\frac{(Y - \mu_X)^2}{2\sigma_X^2}\right\} dY = \frac{1}{2}\left[1 + \text{erf}\left(\frac{X - \mu_X}{\sqrt{2}\,\sigma_X}\right)\right] \tag{B-7}$$

$$\int_{X}^{\infty} \frac{1}{\sqrt{2\pi\sigma_X^2}} \exp\left\{-\frac{(Y - \mu_X)^2}{2\sigma_X^2}\right\} dY = \frac{1}{2}\left[1 - \text{erf}\left(\frac{X - \mu_X}{\sqrt{2}\,\sigma_X}\right)\right] \tag{B-8}$$

$$\int_{X_1}^{X_2} \frac{1}{\sqrt{2\pi\sigma_X^2}} \exp\left\{-\frac{(Y - \mu_X)^2}{2\sigma_X^2}\right\} dY = 1 - \text{erf}\left(\frac{X_1 - \mu_X}{\sqrt{2}\,\sigma_X}\right) + \text{erf}\left(\frac{X_2 - \mu_X}{\sqrt{2}\,\sigma_X}\right) \tag{B-9}$$

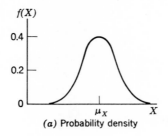

(a) Probability density

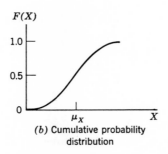

(b) Cumulative probability
distribution

Figure B-1   *Gaussian probability density and distribution functions*

B.2. MOMENTS OF GAUSSIAN DISTRIBUTION

Since the Gaussian probability density is symmetrical about μ_X, the first moment or mean value of X is

$$E(X) = \mu_X \tag{B-10}$$

The variance of X, $E[(X - \mu_X)^2]$, may be found by differentiating

$$\int_{-\infty}^{\infty} \frac{1}{\sqrt{2\pi}} \exp\left\{-\frac{(X - \mu_X)^2}{2\sigma_X^2}\right\} dX = \sigma_X \tag{B-11}$$

with respect to σ_X. Performing the differation yields

$$\int_{-\infty}^{\infty} \frac{(X - \mu_X)^2}{\sqrt{2\pi}\,\sigma_X^3} \exp\left\{-\frac{(X - \mu_X)^2}{2\sigma_X^2}\right\} dX = 1 \tag{B-12}$$

or

$$\int_{-\infty}^{\infty} \frac{(X - \mu_X)^2}{\sqrt{2\pi\sigma_X^2}} \exp\left\{-\frac{(X - \mu_X)^2}{2\sigma_X^2}\right\} dX = \sigma_X^2 \tag{B-13}$$

The integral is the variance of X; hence

$$E\{(X - \mu_X)^2\} = \sigma_X^2 \tag{B-14}$$

The second moment of X is

$$E(X^2) = \sigma_X^2 + [E(X)]^2 = \sigma_X^2 + \mu_X^2 \tag{B-15}$$

In general, for a zero mean Gaussian random variable, the n-th moment is [B-1]

$$E(X^n) = 1 \cdot 3 \cdots (n - 1)\sigma^n \qquad \text{for } n \text{ even} \tag{B-16}$$

$$E(X^n) = 0 \qquad \text{for } n \text{ odd} \tag{B-17}$$

B.3. CHARACTERISTIC FUNCTION OF GAUSSIAN DISTRIBUTION

The characteristic function of a continuous random variable, X, is

$$\Phi(\omega) \equiv E\{\exp(j\omega X)\} \equiv \int_{-\infty}^{\infty} \exp(j\omega X)f(X)\, dX \tag{B-18}$$

For a Gaussian random variable

$$\Phi(\omega) \equiv \int_{-\infty}^{\infty} \frac{\exp(j\omega X)}{\sqrt{2\pi\sigma_X^2}} \exp\left\{-\frac{(X - \mu_X)^2}{2\sigma_X^2}\right\} dX \tag{B-19}$$

Combining the exponential terms and completing the square yield.

$$\Phi(\omega) = \int_{-\infty}^{\infty} \frac{1}{\sqrt{2\pi\sigma_X^2}} \exp\left\{-\frac{[X - (\mu_X + j\omega\sigma_X^2)]}{2\sigma_X^2}\right\} \exp\left\{j\omega\mu_X - \frac{\omega^2\sigma_X^2}{2}\right\} dX \tag{B-20}$$

Since the integral over X is unity,

$$\Phi(\omega) = \exp\left\{j\omega\mu_X - \frac{\omega^2\sigma_X^2}{2}\right\} \tag{B-21}$$

B.4. SUM OR DIFFERENCE OF GAUSSIAN RANDOM VARIABLES

Let $Z = X + Y$ be the sum of two independent Gaussian random variables with parameters μ_X, σ_X and μ_Y, σ_Y, respectively. The characteristic function of Z is the product of the characteristic functions of X and Y.

$$\Phi_Z(\omega) = \Phi_X(\omega)\Phi_Y(\omega) = \exp\left\{j\omega\mu_X - \frac{\omega^2\sigma_X^2}{2}\right\} \exp\left\{j\omega\mu_Y - \frac{\omega^2\sigma_Y^2}{2}\right\}$$

$$= \exp\left\{j\omega(\mu_X + \mu_Y) - \frac{\omega^2}{2}(\sigma_X^2 + \sigma_Y^2)\right\} \tag{B-22}$$

By inspection, $\Phi_Z(\omega)$ is the characteristic function of a Gaussian random variable with mean $\mu_Z = \mu_X + \mu_Y$ and variance $\sigma_Z^2 = \sigma_X^2 + \sigma_Y^2$.

To determine the distribution of the difference of two Gaussian random variables, $Z = X - Y$, let $Y' = -Y$. Then since $E(Y') = -E(Y)$, and the Gaussian distribution is symmetrical, the distribution of Y' is Gaussian with parameters $-\mu_Y$ and σ_Y^2. By the characteristic function method,

$$\Phi_Z(\omega) = \exp\left\{ j(\mu_X - \mu_Y) - \frac{\omega^2}{2}(\sigma_X^2 - \sigma_Y^2) \right\} \tag{B-23}$$

and the distribution of the difference of two Gaussian variables is Gaussian with mean $\mu_Z = \mu_X - \mu_Y$ and variance $\sigma_Z^2 = \sigma_X^2 + \sigma_Y^2$.

B.5. GAUSSIAN PROCESS

A process $Z(t)$ is a Gaussian random process if the random variables $Z(t_1), Z(t_2), \ldots, Z(t_n)$ at time instants $t_1, t_2, \ldots, t_n$ are jointly Gaussian for any n. The process is stationary in the strict sense if its statistics are not affected by a shift in the time origin.

Mean of Gaussian Process

The mean of a Gaussian process is

$$E\{Z(t)\} = \mu_X(t) \tag{B-24}$$

and if the process is stationary, $\mu_X(t) = \mu_X$.

Autocorrelation of Gaussian Process

The autocorrelation function of a stationary Gaussian process is

$$R_Z(\alpha) = E\{Z(t)Z(t + \alpha)\} \tag{B-25}$$

Spectral Density of Gaussian Process

The spectral density of $Z(t)$ is the Fourier transform of the autocorrelation function $R_Z(\alpha)$.

$$G_Z(f) = \int_{-\infty}^{\infty} R_Z(\alpha)e^{-j2\pi f\alpha}\, d\alpha \tag{B-26}$$

B.6. FILTERED GAUSSIAN PROCESS

A filtered Gaussian process, $S(t)$, results when a Gaussian process, $Z(t)$, passes through a linear, time-invariant filter with an impulse response, $h(t)$. By the convolution integral,

$$S(t) = \int_{-\infty}^{\infty} Z(t - \alpha)h(\alpha)\, d\alpha \tag{B-27}$$

Mean of Filtered Gaussian Process

The mean of $S(t)$ is

$$E[S(t)] = \int_{-\infty}^{\infty} E\{Z(t - \alpha)\}h(\alpha)\, d\alpha \tag{B-28}$$

If the process $Z(t)$ is stationary,

$$E[S(t)] = E[Z(t)]\int_{-\infty}^{\infty} h(\alpha)\, d\alpha \tag{B-29}$$

The integral of Equation B-29 is the Fourier transform, $H(j\omega)$, of the filter evaluated at zero frequency

$$H(j\omega) \equiv \int_{-\infty}^{\infty} h(t)e^{-j\omega t}\, d\omega \tag{B-30}$$

Thus the mean of $S(t)$ is

$$E[S(t)] = \mu_X(t)H(0) \tag{B-31}$$

Spectral Density of Filtered Gaussian Process

The power spectral density of the output of a linear filter is equal to the power spectral density of the input multiplied by the absolute value squared of the filter transform. Thus,

$$G_S(f) = |H(j\omega)|^2 G_Z(f) \tag{B-32}$$

Autocorrelation of Filtered Gaussian Process

The autocorrelation of $S(t)$ is the Fourier transform of its spectral density.

$$R_S(\alpha) = \int_{-\infty}^{\infty} |H(j\omega)|^2 G_X(f)e^{j\omega\alpha}\, df \tag{B-33}$$

B.7. NARROW-BAND GAUSSIAN PROCESS [B-2]

The process

$$n(t) = X_c(t)\cos \omega_c t - X_s(t)\sin \omega_c t \tag{B-34}$$

is called a narrow-band Gaussian process under the conditions that $X_c(t)$ and $X_s(t)$ are Gaussian random variables at any time instant and are uncorrelated with the same spectrum $G_X(f)$ which is bandlimited such that $G_X(f) = 0$ for $|f| \geq f_a$. The frequency spectrum of $n(t)$ is then

$$G_n(f) = \tfrac{1}{2}G_X(f - f_c) + \tfrac{1}{2}G_X(f + f_c) \tag{B-35}$$

The process $n(t)$ may be represented as

$$n(t) = V(t)\cos [\omega_c t + \theta(t)] \tag{B-36}$$

where

$$V(t) \equiv \sqrt{X_c^2(t) + X_s^2(t)} \tag{B-37}$$

and

$$\theta(t) = \tan^{-1}\left\{\frac{X_s(t)}{X_c(t)}\right\} \tag{B-38}$$

If $f_a \ll f_c$, the envelope $V(t)$ will vary slowly in the time interval $1/f_c$. Since $X_c(t)$ and $X_s(t)$ are uncorrelated, their joint distribution is the joint Gaussian distribution

$$P(X_c, X_s) = \frac{1}{2\pi\sigma_X^2} \exp\left\{-\frac{(X_c^2 + X_s^2)}{2\sigma_X^2}\right\} \tag{B-39}$$

The joint distribution of V and θ is related to the joint distribution of X_c and X_s by

$$P(V, \theta)\, dV\, d\theta = P(X_c, X_s)\, dX_c\, dX_s \tag{B-40}$$

For the transform given by Equations B-37 and B-38, since $dX_c\, dX_s = V\, dV\, d\theta$, the joint distribution of V and θ becomes

$$P(V, \theta) = \frac{V}{2\pi\sigma_X^2} \exp\left\{\frac{-V^2}{2\sigma_X^2}\right\} \tag{B-41}$$

The distribution of the envelope is found by integrating $P(V, \theta)$ over all values of θ from 0 to 2π. The resulting distribution is called the Rayleigh distribution.

$$P(V) = \frac{V}{\sigma_X^2} \exp\left\{\frac{-V^2}{2\sigma_X^2}\right\} \tag{B-42}$$

Similarly, integrating $P(V, \theta)$ over all positive values of V yields the uniform distribution

$$P(\theta) = \frac{1}{2\pi} \tag{B-43}$$

B.8. NARROW-BAND GAUSSIAN PROCESS PLUS SINE WAVE

If a sine wave $A_c \cos \omega_c t$ is added to the narrowband Gaussian process $n(t)$ given by Equation B-34, the resulting time function may be expressed as

$$Y(t) = [A_c + X_c(t)] \cos \omega_c t - X_s(t) \sin \omega_c t \tag{B-44}$$

The envelope-phase representation of $Y(t)$ is

$$Y(t) = R(t) \cos [\omega_c t + \psi(t)] \tag{B-45}$$

where

$$R(t) \equiv \sqrt{[X_c(t) + A_c]^2 + X_s^2(t)} \qquad (B\text{-}46)$$

and

$$\psi(t) \equiv \tan^{-1}\left[\frac{X_s(t)}{X_c(t) + A_c}\right] \qquad (B\text{-}47)$$

The random variable $X_c'(t) \equiv A_c + X_c(t)$ is Gaussianly distributed with mean A_c. Hence

$$P(X_c', X_s) = \frac{1}{2\pi\sigma_X^2} \exp\left\{-\frac{[(X_c' - A_c)^2 + X_s^2]}{2\sigma_X^2}\right\} \qquad (B\text{-}48)$$

The joint distribution $P(R, \psi)$ may be found for the transform given by Equations B-46 and B-47 along with the substitution $X_c' = R\cos\psi$.

$$P(R, \psi) = \frac{R}{2\pi\sigma_X^2} \exp\left\{-\frac{[R^2 + A_c^2 - 2A_cR\cos\psi]}{2\sigma_X^2}\right\} \qquad (B\text{-}49)$$

Integrating $P(R, \psi)$ over ψ from 0 to 2π yields the distribution of the envelope, R.

$$P(R) = \frac{R}{2\pi\sigma_X^2} \exp\left\{-\frac{(R^2 + A_c^2)}{2\sigma_X^2}\right\} \int_0^{2\pi} \exp\left\{\frac{A_cR\cos\psi}{\sigma_X^2}\right\} d\psi \qquad (B\text{-}50)$$

The integral may be written in terms of the modified Bessel function of order zero.

$$I_0(W) \equiv \frac{1}{2\pi} \int_0^{2\pi} \exp\{W\cos\psi\} d\psi \qquad (B\text{-}51)$$

Hence the envelope distribution is

$$P(R) = \frac{R}{\sigma_X^2} \exp\left\{-\frac{(R^2 + A_c^2)}{2\sigma_X^2}\right\} I_0\left[\frac{A_cR}{\sigma_X^2}\right] \qquad (B\text{-}52)$$

For $A_c = 0$, $P(R)$ reduces to the Rayleigh distribution; for A_c large, $P(R)$ approaches a Gaussian distribution.

REFERENCES

B-1. Papoulis, A. *Probability, Random Variables, and Stochastic Processes*. McGraw-Hill, New York, 1965.

B-2. Davenport, W. B. Jr. and Root, W. L. *An Introduction to the Theory of Random Signals and Noise*. McGraw-Hill, New York, 1958.

POISSON PROCESS

C.1. POISSON RANDOM VARIABLE

A discrete random variable X is Poisson distributed with parameter λ if its probability of taking integral values is

$$P(X = k) = \frac{\exp(-\lambda)(\lambda)^k}{k!} \qquad k = 0, 1, 2, \ldots. \tag{C-1}$$

The probability density and cumulative distribution functions, as illustrated in Figure C-1, are

$$f(X) = \sum_{k=0}^{\infty} \frac{\exp(-\lambda)(\lambda)^k}{k!} \, \delta(X - k) \tag{C-2}$$

and

$$F(X) = \sum_{k=0}^{\infty} \frac{\exp(-\lambda)(\lambda)^k}{k!} \, U(X - k) \tag{C-3}$$

C.2. MOMENTS OF POISSON DISTRIBUTION

The first moment or mean value of X is

$$E(X) = \sum_{k=1}^{\infty} k \, \frac{\exp(-\lambda)(\lambda)^k}{k!} = \lambda \sum_{k=1}^{\infty} \frac{\exp(-\lambda)(\lambda)^{k-1}}{(k-1)!} \tag{C-4}$$

Since the latter summation is unity,

$$E(X) = \lambda \tag{C-5}$$

The second moment of X is

$$E(X^2) = \sum_{k=1}^{\infty} k^2 \, \frac{\exp(-\lambda)(\lambda)^k}{k!} = \lambda \sum_{k=1}^{\infty} k \, \frac{\exp(\lambda)(\lambda)^{k-1}}{(k-1)!} \tag{C-6}$$

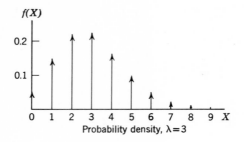

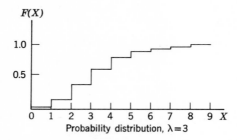

Figure C-1   *Poisson probability density and distribution functions*

Letting $j = k - 1$, the summation on the right reduces to

$$E(X^2) = \lambda \sum_{j=0}^{\infty} j \frac{\exp(-\lambda)(\lambda)^j}{j!} + \lambda \sum_{j=0}^{\infty} \frac{\exp(-\lambda)(\lambda)^j}{j!} \qquad (C-7)$$

The first summation is the mean λ and the second summation is unity, yielding

$$E(X^2) = \lambda^2 + \lambda \qquad (C-8)$$

The variance is by definition

$$\sigma_X^2 = E(X^2) - [E(X)]^2 \qquad (C-9)$$

and hence

$$\sigma_X^2 = \lambda \qquad (C-10)$$

Thus, the mean and variance of the Poisson random variable are identical.

C.3. CHARACTERISTIC FUNCTION OF POISSON DISTRIBUTION

The characteristic function of a discrete random variable is

$$\Phi(\omega) = \sum_{k=0}^{\infty} \exp(j\omega k)P(X = k) \qquad (C-11)$$

For a Poisson random variable

$$\Phi(\omega) = \sum_{k=0}^{\infty} \frac{\exp(j\omega k)\exp(-\lambda)(\lambda)^k}{k!} = \exp(-\lambda) \sum_{k=0}^{\infty} \frac{[\lambda\exp(j\omega)]^k}{k!} \quad \text{(C-12)}$$

The series on the right is the exponential series representation of $\lambda\exp(j\omega)$; hence

$$\Phi(\omega) = \exp\{\lambda[\exp(j\omega) - 1]\} \quad \text{(C-13)}$$

C.4. SUM OF POISSON VARIABLES

Let $Z = X + Y$ be the sum of two independent Poisson random variables with parameters λ_X and λ_Y, respectively. The characteristic function of Z is the product of the characteristic functions of X and Y.

$$\Phi_Z(\omega) = \Phi_X(\omega)\Phi_Y(\omega) = \exp\{[\lambda_X + \lambda_Y][\exp(j\omega) - 1]\} \quad \text{(C-14)}$$

Thus, $\Phi_Z(\omega)$ is the characteristic function of a Poisson random variable with parameter $\lambda_X + \lambda_Y$, and the sum of Poisson random variables is a Poisson random variable.

$$P(Z = j) = \frac{\exp[-(\lambda_X + \lambda_Y)](\lambda_X + \lambda_Y)^j}{j!} \quad \text{(C-15)}$$

C.5. DIFFERENCE OF POISSON VARIABLES

Let $Z = X - Y$ be the difference of two independent Poisson random variables with parameters λ_X and λ_Y. The distributions of Z are

$$P(Z = j) = \sum_{m=0}^{\infty} P(X = j + m)P(Y = m) \quad \text{for } j \geq 0 \quad \text{(C-16)}$$

and

$$P(Z = j) = \sum_{m=0}^{\infty} P(X = m)P(Y = j + m) \quad \text{for } j < 0 \quad \text{(C-17)}$$

For $j \geq 0$, substituting for the Poisson distribution yields

$$P(Z = j) = \sum_{m=0}^{\infty} \frac{\exp(-\lambda_X)(\lambda_X)^{j+m}}{(j+m)!} \frac{\exp(-\lambda_Y)(\lambda_Y)^m}{m!} \quad \text{(C-18)}$$

The modified Bessel function of the first kind of order j is

$$I_j(\beta) = \sum_{m=0}^{\infty} \frac{(\beta/2)^{2m+j}}{m!\,(m+j)!} \quad \text{(C-19)}$$

Letting $\beta = \sqrt{2\lambda_X\lambda_Y}$ leads to the expression

$$P(Z = j) = \exp\left[-(\lambda_X + \lambda_Y)\right]\left(\frac{\lambda_X}{\lambda_Y}\right)^{j/2}I_j[2\sqrt{\lambda_X\lambda_Y}] \qquad \text{for } j \geq 0 \quad (C\text{-}20)$$

Similarly, for $j < 0$,

$$P(Z = j) = \sum_{m=0}^{\infty} \frac{\exp(\lambda_X)(\lambda_X)^m}{m!} \frac{\exp(\lambda_Y)(\lambda_Y)^{j+m}}{(j+m)!} \qquad (C\text{-}21)$$

which can be written as

$$P(Z = j) = \left(\frac{\lambda_Y}{\lambda_X}\right)^{j/2} \exp\left[-(\lambda_X + \lambda_Y)\right]I_j[2\sqrt{\lambda_X\lambda_Y}] \qquad \text{for } j < 0 \quad (C\text{-}22)$$

Combining Equations C-20 and C-22 yields the probability distribution of Z.

$$P(Z = j) = \exp\left[-(\lambda_X + \lambda_Y)\right]\left(\frac{\lambda_Y}{\lambda_X}\right)^{-j/2}I_{|j|}[2\sqrt{\lambda_X\lambda_Y}] \qquad (C\text{-}23)$$

C.6. POISSON PROCESS [C-1]

A stochastic process

$$Z(t) = C\sum_{n=1}^{k} \delta(t - t_n) \qquad (C\text{-}24)$$

composed of a sequence of impulses occurring at times t_n in a time period $-\tau/2$ to $\tau/2$ and multiplied by a constant C is a Poisson process if the probability that the number, X, of impulses occurring in the time period τ is an integer is

$$P\{X(t) = k\} = \frac{\exp(-\lambda t)(\lambda t)^k}{k!} \qquad (C\text{-}25)$$

where λ equals the constant average time rate of occurrence of impulses. The process $Z(t)$, as shown in Figure C-2, is related to $X(t)$ by

$$Z(t) = \frac{dX(t)}{dt} \qquad (C\text{-}26)$$

The probability density of $Z(t)$ is

$$P\{Z(t)\} = P\{t_1, t_2, \ldots, t_K \mid k\}P\{X(t) = k\} \qquad (C\text{-}27)$$

which is not Poisson. If the times of occurences of impulses are independent of one another and independent of the total number of impulses,

$$P\{Z(t)\} = \frac{1}{\tau^k} P\{X(t) = k\} \qquad (C\text{-}28)$$

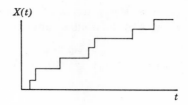

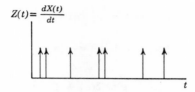

Figure C-2   *Poisson process*

Mean of Poisson Process

The first moment of $Z(t)$ is

$$E\{Z(t)\} = E\left\{\frac{dx\,(t)}{dt}\right\} = \frac{d}{dt}\,E\{X(t)\} = \frac{d\,(\lambda t)}{dt} \tag{C-29}$$

and hence

$$E\{Z(t)\} = \lambda \tag{C-30}$$

Autocorrelation of Poisson Process

The autocorrelation function of a Poisson process is

$$R_Z(\alpha) = E\{Z(t)Z(t + \alpha)\} \tag{C-31}$$

Averaging over the impulse times $t_1, t_2, \ldots, t_k$ and the random variable k yields

$$R_Z(\alpha) = \int_{-\tau/2}^{\tau/2} \cdots \int_{-\tau/2}^{\tau/2} \sum_{k=0}^{\infty} \left[C\sum_{n=1}^{k}\delta(t - t_n)\right]\left[C\sum_{m=1}^{\infty}\delta(t - \alpha - t_m)\right]$$

$$\times \frac{1}{\tau^k}\,P\{X(t) = k\}\,dt_1 \cdots d_{t_k} \tag{C-32}$$

Changing the order of integration and summation gives [C-2]

$$R_Z(\alpha) = \sum_{k=0}^{\infty} P\{X(t) = k\}\left\{\sum_{n=1}^{k}\int_{-\tau/2}^{\tau/2}\frac{C^2}{\tau}\,\delta(t)\,\delta(t + \alpha)\,dt\right.$$

$$+ \underbrace{\sum_{n=1}^{k}\sum_{m=1}^{k}}_{m \neq m}\int_{-\tau/2}^{\tau/2}\frac{C}{\tau}\,\delta(t - t_n)\,dt_n\int_{-\tau/2}^{\tau/2}\frac{C}{\tau}\,\delta(t + \alpha - t_m)\,dt_m\right\} \tag{C-33}$$

The resulting integration over the k terms of the first summation in the brackets and the $k^2 - k$ terms of the double summation gives

$$R_Z(\alpha) = \sum_{k=0}^{\infty} P\{X(t) = k\}\left[\frac{kC^2}{\tau}\,\delta(\alpha) + (k^2 - k)\frac{C^2}{\tau^2}\right] \qquad \text{(C-34)}$$

or

$$R_Z(\alpha) = \frac{C^2}{\tau}\,E\{X(t)\}\,\delta(\alpha) + \frac{C^2}{\tau^2}\,E\{[X(t)]^2\} - \frac{C^2}{\tau^2}\,E\{X(t)\} \qquad \text{(C-35)}$$

Since $E\{X(t)\} = \lambda\tau$ and $E\{[X(t)]^2\} = (\lambda^2\tau^2 + \lambda\tau)$, the autocorrelation of $Z(t)$ is

$$R_Z(\alpha) = C^2\lambda\,\delta(\alpha) + C^2\lambda^2 \qquad \text{(C-36)}$$

Spectral Density of Poisson Process

The spectral density of $Z(t)$ is the Fourier transform of the autocorrelation function $R_Z(\alpha)$.

$$G_Z(f) = \int_{-\infty}^{\infty} C^2[\lambda\,\delta(\alpha) + \lambda^2]\exp\left(-j2\pi f\alpha\right)d\alpha \qquad \text{(C-37)}$$

and

$$G_Z(f) = C^2\lambda + C^2\lambda^2\,\delta(f) \qquad \text{(C-38)}$$

C.7. FILTERED POISSON PROCESS

A filtered Poisson process

$$S(t) = \sum_{n=1}^{k} h(t - t_n) \qquad \text{(C-39)}$$

results when the Poisson process, $Z(t)$, passes through a linear, time-invariant filter with an impulse response of $h(t)$ (Figure C-3). By the convolution integral

$$S(t) = \int_{-\infty}^{\infty} Z(t - \alpha)h(\alpha)\,d\alpha \qquad \text{(C-40)}$$

Mean of Filtered Poisson Process

The mean of $S(t)$ is

$$E\{S(t)\} = \int_{-\infty}^{\infty} E\{Z(t - \alpha)\}h(\alpha)\,d\alpha \qquad \text{(C-41)}$$

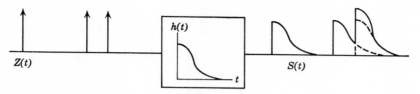

Figure C-3 *Filtered poisson process*

If the process $Z(t)$ is stationary

$$E\{S(t)\} = E\{Z(t)\} \int_{-\infty}^{\infty} h(\alpha) \, d\alpha \qquad \text{(C-42)}$$

The integral of Equation C-42 is the Fourier transform, $H(j\omega)$, of the filter evaluated at zero frequency

$$H(j\omega) \equiv \int_{-\infty}^{\infty} h(t)e^{-j\omega t} \, dt \qquad \text{(C-43)}$$

Thus the mean of $S(t)$ is

$$E\{S(t)\} = \lambda H(0) \qquad \text{(C-44)}$$

Spectral Density of Filtered Poisson Process

The power spectral density of the output of a linear filter is equal to the power spectral density of the input multiplied by the absolute value squared of the filter transform. Thus

$$G_S(f) = |H(j\omega)|^2 G_Z(f) \qquad \text{(C-45)}$$

and

$$G_S(f) = C^2\lambda|H(j\omega)|^2 + C^2\lambda^2[H(0)]^2 \, \delta(f) \qquad \text{(C-46)}$$

Autocorrelation of Filtered Poisson Process

The autocorrelation of $S(t)$ is the inverse Fourier transform of its power spectral density.

$$R_S(\alpha) = \int_{-\infty}^{\infty} \{C^2\lambda|H(j\omega)|^2 + C^2\lambda^2[H(0)]^2 \, \delta(f)\} \exp\{j\omega\alpha\} \, df \qquad \text{(C-47)}$$

Since $|H(j\omega)|^2 = H(j\omega)H(-j\omega)$, the autocorrelation is

$$R_S(\alpha) = C^2\lambda \int_{-\infty}^{\infty} H(j\omega)H(-j\omega) \exp\{j\omega\alpha\} \, df + C^2\lambda^2[H(0)]^2 \qquad \text{(C-48)}$$

The integral is equivalent to a convolution of $h(t)$ with its time inverse $h(t)$ Hence

$$R_S(\alpha) = C^2\lambda \int_{-\infty}^{\infty} h(\alpha + \beta)h(\beta) \, d\beta + C^2\lambda^2[H(0)]^2 \qquad \text{(C-49)}$$

REFERENCES

C-1. Papoulis, A. *Probability, Random Variables, and Stochastic Processes*. McGraw-Hill, New York, 1965.

C-2. Davenport, W. B. Jr. and Root, W. L. *An Introduction to the Theory of Random Signals and Noise*. McGraw-Hill, New York, 1958.

appendix D

LIST OF SYMBOLS

A_c = laser carrier amplitude
A_o = local oscillator amplitude
B_i = optical input filter bandwidth in frequency units
B_O = output filter bandwidth
B_{IF} = intermediate frequency filter bandwidth

$\dfrac{C}{N}$ = power carrier-to-noise ratio

$C(t)$ = instantaneous carrier intensity
$C_M(t)$ = instantaneous modulated carrier intensity
C_n = atmospheric turbulence structure constant
c = velocity of light in free space
$\mathcal{D}$ = detector conversion factor
D = photodetector detectivity
d_P = photodetector diameter
d_R = receiver antenna diameter
d_T = transmitter antenna diameter
d_S = diameter of background radiation source
d_B = laser beam diameter
$E(t)$ = instantaneous carrier electric field
$E_M(t)$ = instantaneous modulated carrier electric field
E_g = energy gap energy
f_c = transmission frequency
f_o = local oscillator frequency
f_m = modulation sinusoid frequency
F = focal length
G = photodetector current gain
$G_{i_P}(f)$ = photodetector shot noise power spectral density in current units
$G_{i_D}(f)$ = dark current shot noise power spectral density in current units
$G_{i_B}(f)$ = background radiation shot noise power spectral density in current units
$G_{i_S}(f)$ = laser carrier shot noise power spectral density in current units

G_{v_T} = thermal noise power spectral density at optical receiver output in voltage units

h = Planck's constant

$\mathscr{H}(\lambda)$ = background radiation spectral irradiance in wavelength units

I_P = average photodetector current

I_B = average photodetector current due to background radiation

I_D = average photodetector current due to dark current

i_P = instantaneous photodetector current

i_H = instantaneous photodetector current due to background radiation and dark current

i_F = filtered photodetector current

i_{IF} = intermediate frequency signal current

k = Boltzmann's constant

$\mathscr{L}$ = laser carrier polarization matrix

l = atmospheric turbulence inhomogeneity dimension

$L(t)$ = instantaneous local oscillator electric field

$M(t)$ = electrical information signal

M_{AM} = AM modulation index

M_{FM} = FM modulation index

M_{IM} = IM modulation index

$\mathscr{N}(\lambda)$ = background radiation spectral radiance in wavelength units

N_H = shot noise power at optical receiver output

N_T = thermal noise power at optical receiver output

$P_e{}^B$ = probability of detection error per bit

$P_e{}^P$ = probability of detection error per sample

P_{SN} = probability that photodetector signal plus noise exceeds detection threshold

P_N = probability that photodetector noise exceeds detection threshold

$P(i_S)$ = probability density of photodetector current due to a laser carrier

$P(i_T)$ = probability density of thermal noise current

$P(i_F)$ = probability density of filtered photodetector current

$P[U_{S,\tau} = k]$ = probability distribution of number of laser carrier photoelectron emissions in time interval τ

$P[U_{B,\tau} = k]$ = probability distribution of number of background radiation photoelectron emissions in time interval τ

$P[U_{D,\tau} = k]$ = probability distribution of number of dark current photoelectron emissions in time interval τ

$P[U_{H,\tau} = k]$ = probability distribution of number of dark current and background radiation photoelectron emissions in time interval τ

P_C = average carrier power at photodetector surface

P_B = average background radiation power incident upon detector surface

P_L = laser transmitter power
P_O = local oscillator power
P_A = antenna output power
q = electronic charge
$\mathscr{Q}(\lambda)$ = background radiation photon spectral radiance in wavelength units
R_L = optical receiver load resistor
R = transmission range
r_o = heterodyne detection phase coherence dimension
S = signal power
$\dfrac{S}{N}$ = power signal-to-noise ratio

T = temperature
T_S = temperature of background radiation source
$\mathscr{W}(\lambda)$ = background radiation spectral radiant emittance in wavelength units
$\mathscr{W}(f)$ = background radiation spectral radiant emittance in frequency units
α_a = atmospheric attenuation coefficient
α_b = atmospheric absorption coefficient
α_s = atmospheric scattering coefficient
α_{SR} = atmospheric Rayleigh scattering coefficient
α_{SM} = atmospheric Mie scattering coefficient
Γ = phase retardation between orthogonal electric field components
η = quantum efficiency
θ_T = transmitter antenna beamwidth
θ_R = receiver antenna field of view
λ_c = transmission wavelength
λ_i = optical input filter bandwidth in wavelength units
$\mu_{R,\tau}$ = average number of photoelectrons emitted in time interval τ due to a general optical source of radiation
$\mu_{D,\tau}$ = average number of photoelectrons emitted in time interval τ due to dark current emissions
$\mu_{S,\tau}$ = average number of photoelectrons emitted in time interval τ due to a laser carrier
$\mu_{B,\tau}$ = average number of photoelectrons emitted in time interval τ due to background radiation
$\mu_{H,\tau}$ = average number of photoelectrons emitted in time interval τ due to background radiation and dark current
ρ_0 = lateral phase coherence length
$\sigma_\Phi{}^2(\rho)$ = variance of phase fluctuations between points in a laser beam separated by ρ
τ_B = bit period
τ_P = sample period

τ_C = coherence time

τ_a = atmospheric transmissivity

τ_t = transmitter system transmissivity

τ_r = receiver system transmissivity

τ_b = atmospheric absorption transmissivity

τ_s = atmospheric scattering transmissivity

Φ_c = carrier phase angle

Φ_o = local oscillator phase angle

$\psi\left(\dfrac{d_R}{r_o}\right) = $ heterodyne detection signal-to-noise ratio degradation factor due to atmospheric turbulence

Ω_R = solid angle of receiver field of view

Ω_S = solid angle subtended by the source at the receiver

ω_c = carrier angular frequency

ω_d = peak angular frequency deviation

ω_o = local oscillator angular frequency

ω_m = modulation sinusoid angular frequency

INDEX